The Really Hard Problem

The Really Hard Problem

Meaning in a Material World

Owen Flanagan

A Bradford Book
The MIT Press
Cambridge, Massachusetts
London, England

For information on quantity discounts, email special_sales@mitpress.mit.edu.

Set in Stone Serif and Stone Sans on 3B2 by Asco Typesetters, Hong Kong. Printed and bound in the United States of America.

Library of Congress Cataloging-in-Publication Data

Flanagan, Owen J.
The really hard problem : meaning in a material world / Owen Flanagan.
p. cm.
"A Bradford book."
Includes bibliographical references and index.
ISBN 978-0-262-06264-0 (hardcover : alk. paper)
1. Meaning (Philosophy) 2. Cognitive psychology. 3. Materialism—Psychological aspects. I. Title.
B105.M4F53 2007
121′.68—dc22 2007002664

10 9 8 7 6 5 4 3 2 1

to Güven Güzeldere and David Wong, dear friends and spiritually advanced natural philosophers

Contents

Acknowledgments

The occasion for this book was an invitation to give the Templeton Research Lectures at the University of Southern California. I was invited—with a year's preparation—to talk about how things, considered in the broadest possible sense, hang together (if they do) in the broadest possible sense. Specifically, I was asked—or so I interpreted the invitation—to focus on the implications of mind science for our conception of ourselves. The original title of the series of lectures that became this book was Human Flourishing in the Age of Mind Science. I gave the lectures in Los Angeles over a glorious two-week period in February of 2006. I could not have felt more welcomed than by the USC group that sponsored the lectures and their wonderful support staff. Firdaus Udwadia and Nicolas Lori seemed always at my side, thankfully sharing my view that the scientific image need not be understood as disenchanting. Several graduate seminars at Duke University provided chances to try out some of the material. I am especially grateful to Joost Bosland for reading the lectures carefully and designing a wonderful PowerPoint presentation with artwork beyond my dreams. Jeremy Evans, Sahar Akthar, Russ Powell, and Robert Williams stand out for making me think very hard about some parts of my overall line of argument. Then there are David Wong and Hagop Sarkissian. I trusted them with most every word you have before you and they gave wonderful, tough critical responses.

There is also the "Mind and Life" crowd. We are a group of philosophers and scientists who have been involved in discussions with the Dalai Lama about science and spirituality in many meetings in India and America over the last decade. I am grateful to them all. Alan Wallace and Rob Hogendoorn stand out as especially helpful critics. Finally, as always, there are public lectures. I owe gratitude to the audiences in Los Angeles who listened to and commented on my lectures. Then there were chances at the University of Hawaii, in Santa Barbara, at Bowdoin College, at Duke University, at the Esalen Institute (thanks especially to Mike Murphy and the human potential movement), and at Columbia University to test and re-test these ideas. Thanks to Chris Kelly, Bob Pollack, and Bob Thurman for that amazing Mind and Reality conference in New York in March of 2006.

My colleagues and students at Duke—not only the ones already mentioned, but the whole department—have helped me to think more clearly about science and life's meanings. Alex Rosenberg, Güven Güzeldere, and David Wong made me think especially hard about my upbeat arguments for finding meaning in a material world. I cannot, of course, satisfy myself, my critics, or my friends that I provide an answer to the really hard problem of the meaning(s) of life. I hope, out of gratitude to my ancestors, in honor of my family, my friends, and my critics, and in service to the well-being of all sentient beings, to have said something useful, something in the right direction, something that might matter to contributing to the realization of what is true, good, and beautiful.

Introduction

Within mind science, "the hard problem" is to explain how mind is possible in a material world. How could the amazing private world of my consciousness emerge out of neuronal activity? This problem is hard. But it is even harder to explain how meaning is possible in this material world. Nearly everyone accepts that consciousness exists. Many wonder whether meaning does, even could, exist. Consciousness is. It happens, it is there. It flows like a stream while I live, and how it flows, how it connects to itself, is what makes me who I am. Meaning, if there is such a thing, is a matter of whether and how things add up in the greater scheme of things. Meaning, unlike consciousness, is not simply a puzzling feature of the way things are. Whether there is or can be such a thing as meaning is a more complicated matter than what there is. Unlike consciousness, meaning isn't a matter of what there is or isn't. Meaning, if there is such a thing, involves more than what there is. Minimally it involves a truthful assessment of what living a finite human life adds up to.

How is consciousness possible? How does subjectivity emerge from objective biological features of the nervous system? What is the function of consciousness? What does it do, and how much? How, when, and why did consciousness evolve in certain animal lineages? What does living as a self-aware social mammal mean or add up to? How does living a conscious embodied life matter, add up to anything—anything at all?

I have come to think that how to make sense of living meaningfully is the hardest question. Consciousness exists. There is no doubt that we are conscious creatures. Indeed, consciousness has the effect in the case of humans of enabling us to ask such questions as "What makes life meaningful?" "What does my life, or any human life for that matter, add up to?"

and "Why and how, in the greater scheme of things, does any human life matter?"

Consciousness exists, and if we accept Darwin's theory it probably serves a biological function. But whether meaning exists is controversial. We tell stories about what it is to live a meaningful life. But it is not clear that any of these stories give us insight, let alone an answer, to the question of what a truly meaningful life is or might be. We can imagine respectable answers to the first two questions emerging from the mind sciences and evolutionary biology, respectively. The question of meaning, if it has a good answer, seems to require more resources than these sciences. In fact, many will say that the mind sciences and evolutionary biology are part of the problem, not part of the solution to the problem of meaning. These sciences presuppose that we are finite biological beings living in a material world. If there is meaning, it must be a kind suited to us, a certain kind of conscious mammal who lives three score years and ten and then is gone. Gone forever. In order to address the really hard problem, let alone begin to answer it, I find it necessary to widen the scope of disciplines involved in the inquiry to include not only all mind sciences and evolutionary biology but also Western and Eastern philosophy, political theory, the history of religion, and what is nowadays called "positive psychology." Anthropology, sociology, and economics are also major contributors to this exercise in eudaimonics, the attempt to say something naturalistic and systematic about what makes for human flourishing and that gives life meaning—if, that is, anything does.

We are conscious social animals. There is little doubt about that. How consciousness emerges from our biology is puzzling. But the really hard problem—in the sense that it is existentially pressing—is that it might be true that we are conscious beings who seek to live meaningfully, but that there is nothing that could make this aspiration real, nothing more than a wish that comes with being a conscious social animal. Maybe worrying about "real meaning" is the source of the angst. Perhaps we bring to the table fantasies rather than realistic expectations about what "real" or "genuine" meaning would be. It is hard to know.

How does a naturalist make sense of the meaning, magic, and mystery of life? How does one say truthful and enchanting things about being human? It is not clear. Here I make an attempt to explain how we can make sense and meaning of our lives given that we are material beings living in a material world. The picture I propose is naturalistic and enchanting. Or so I hope.

When I was a wee boy abounding with skeptical religious thoughts, I comforted myself with the notion that God—if he existed—would not punish me for seeking the truth. I no longer believe in God, at least not the kind of God that I was taught to believe in. But I maintain the idea that we humans should not suffer from the truth. Some friends say I seek a way to make the truth consoling, to make a bitter pill palatable. Perhaps. There are worse things than being truthful and consoling. But I don't like the "bitter pill" analogy. Bitter taste is relational; that is, a bitter taste is not in the world. In that respect taste is like meaning. It may be that we are hard-wired to find certain flavors bitter. The analogy breaks down because there are no brain buds—like taste buds—that are automatically set to find certain truths about our predicament depressing or disenchanting. So I say. We can adopt different legitimate attitudes toward the truth about our nature and our predicament. I recommend optimistic realism. Joyful optimistic realism. Life can be precious and funny. And one doesn't need to embrace fantastical stories—unbecoming to historically mature beings—about our nature and prospects to make it so.

All Saints' Day (November 1), 2006

The Really Hard Problem

1 Meaningful and Enchanted Lives: A Threat from the Human Sciences?

Meaningful Lives and the Scientific Image of Persons

What sense can be made of my wish to live in a genuinely meaningfully way, to live a life that really matters, that makes a positive and lasting contribution, if my life is exhausted by my prospects as a finite material being living in a material world? To be sure, I, like all other humans, wish to flourish, to be blessed with happiness, to achieve eudaimonia—to be "a happy spirit." So what? How could eudaimonia really be in store for a short-lived piece of organized muscle and tissue that happens to be aware of its predicament and wishes to flourish? Suppose I am lucky enough to live a blessed life in the sense that I feel happy, think I live well and experience self-respect. What does this add up to? How does it matter, if when I die I am gone forever? I aim to address these questions here.

There are surprisingly favorable prospects for a type of empirical-normative inquiry suited for our kind of animal that explains what genuine flourishing is, how it is possible for creatures like us, and what methods are available to achieve it. I call this empirical-normative inquiry into the nature, causes, and conditions of human flourishing *eudaimonistic scientia*. 'Eudaimonia' is a Greek word that means "flourishing"; 'scientia' is a Latin word that means "knowledge."[1] Despite having the odd property of crossing two dead languages, 'eudaimonistic scientia' captures what I want it to capture.

Eudaimonistic scientia—eudaimonics, for short—is based on 2,500 years of observation and study of our kind of being. The aim of eudaimonics is human flourishing (and the flourishing of other sentient beings), and any and all reliable means to the production of flourishing are in its arsenal.[2] "Project Eudaimonia" is how we fans advertise our efforts. The important thing is that eudaimonics is empirical, or, better, epistemologically responsible—all claims about the nature, causes, and conditions of flourishing are to be based on reasoning about the evidence, historical and contemporary, as to what flourishing is (including its varieties), and what its causes and constituents are. Although eudaimonics is not itself a science in the modern sense, it involves systematic philosophical theorizing that is continuous with science and which therefore takes the picture of persons that science engenders seriously. Eudaimonics is one way for the naturalist to respond to those who say that science in general and the human sciences in particular disenchant the world in the sense that they take away all the meaning and significance that magical, wishful stories once provided.

Ever since Darwin, we have been asked to re-conceive our nature. We are not embodied souls, nor are we bodies with autonomous Cartesian minds. We are animals. The fact that we are animals does not reveal who and what we are or what our prospects are as human animals. It serves primarily as a constraint on how we ought to think about our *Dasein*, our being in the world. Whatever we are, or turn out to be, cannot depend on possessing any capacities that are not natural for fully embodied beings. This, happily, is compatible with possessing amazing and previously unseen natural abilities.

'Naturalism' names a modest position. It serves primarily to mark my orientation off from non-naturalistic and especially supernaturalistic views. In metaphysics and in philosophy of mind, the objectionable views are impressed by the powerful idea of the *scala natura*, "the Great Chain of Being." Humans sit uniquely poised between minerals, plants, and animals on one side and spirits—angels and God—on the other side, and thus partake of both natures. We are part body, part non-physical mind or soul. Mind operates according to *sui generis* principles that allow circumvention of ordinary physical laws, including dying and being dead. *Res cogitans*—mind conceived as non-physical, as immaterial "thinking stuff"—allows for but doesn't entail eternal life.

This sort of non-naturalist view provides a clear contrastive space in which to get a modest naturalism up and running. Naturalism is impressed by the causal explanatory power of the sciences. Science typically denies the truth—or at least the testability—of theories that invoke non-natural, occult, or supernatural causes or forces.

Conceived this way, philosophical naturalism reins in temptations to revert to dualistic and/or supernaturalistic ways of speaking and thinking about human nature. And it does so for principled reasons. Non-naturalistic ways of conceiving of persons face insurmountable problems, for example, explaining how it is even possible for mind and body to causally interact. Naturalistic conceptions of persons lead to progressive theorizing; non-naturalistic ones do not.[3]

A broad philosophical naturalism can accommodate our unusual nature as social animals that both discover and make meaning. If this is right, there is nothing inherently disturbing or disenchanting about the naturalistic picture of human being. We ought to beware scientism, but the scientific image of persons need not make us weak in the knees. Even if I am an animal, even if at the end of the day I am dead and gone for good, I still make a difference, good or bad. Why? Because I exist. Each existing thing makes a difference to how things go—a small difference, but a difference. It would be nice to know, given that I care, how to contribute a bit to the accumulation of good effects, or ones I hope will be positive. Eudaimonics is intended to provide an empirical framework for thinking about human flourishing.[4]

Project Eudaimonia

Darwin's theory is the cornerstone for a fully naturalistic theory of persons. The theory of evolution by natural selection provides prospects for philosophical unification of all the sciences that pertain to human being. Because we are animals, biochemistry, genetics, and neuroscience allow us to see heretofore unseen aspects of ourselves more deeply and truthfully. The so-called moral sciences or *Geisteswissenschaften* (literally sciences of the spirit) are re-conceived. Anthropology, economics, political science, and sociology study the thinking and being of social animals, not collections of radically autonomous Cartesian agents, not of beings running on *Geist*—on spiritual fuel in the spooky sense. The unification of the sciences that study persons[5] is made possible by the insight that all these sciences are all engaged in studying various aspects of the thinking and being of a certain very smart species of social mammal.

In my experience, most people don't like it when I press this idea, the idea that we are animals, although most will admit to finding themselves perplexed. On the one hand, many see that this picture of persons is required by mature acceptance of the message of a hugely successful scientific synthesis that has been progressing for 150 years. On the other hand, the naturalistic way of conceiving of persons feels reductive and disenchanting, especially if it is stated or implied that our prospects are exhausted by whatever ends there are for fully material beings. The scientific image of persons drains the cup that sustains us of whatever it is that could conceivably give human life real depth, texture, and meaning. If it is true that we are material beings living in a material world, especially if all our prospects are exhausted by our nature as finite animals, then that is depressing. And if you believe it, even if it can be shown to be true, keep it quiet.

One question that needs sustained exploration is this: What does it mean to be a material being living in a material world? What does it mean to be a conscious being if at the end of the day we are just a temporarily organized system of particles, or, as seen at another level, just a hunk of meat? Some say or worry that it means that nothing is as it seems, and life really is meaningless. Others, like me, think that living meaningfully continues more or less as before with a promising potential upside that paradoxically comes from accepting naturalism. If one adopts the perspective of the philosophical naturalist and engages in realistic empirical appraisal of our natures and prospects, we have chances for learning what methods might reliably contribute to human flourishing. This is eudaimonics.

Eudaimonics, as I conceive it and depict it in what follows, provides a framework for thinking in a unified way about philosophical psychology, moral and political philosophy, neuroethics, neuroeconomics, and positive psychology, as well as about transformative mindfulness practices that have their original home in non-theistic spiritual traditions such as Buddhism, Aristotelianism, and Stoicism. The latter disciplines, inquiries, sciences, and spiritual practices, insofar (and this is not their only aim) as they seek to understand the nature, causes, and constituents of well-being and to advance flourishing, are parts of Project Eudaimonia. Eudaimonics is the activity of systematically gathering what is known about these three components of well-being and attempting to engender as much flourishing as is possible.

The Philosopher's Vocation

In 1960, Wilfrid Sellars began a famous pair of lectures as follows:

> The aim of philosophy, abstractly formulated, is to understand how things in the broadest possible sense of the term hang together in the broadest possible sense of the term. Under 'things in the broadest possible sense' I include such radically different items as not only 'cabbages and kings', but numbers and duties, possibilities and finger snaps, aesthetic experience and death. To achieve success in philosophy would be, to use a contemporary turn of phrase, to 'know one's way around' with respect to all these things, not in that unreflective way the centipede of the story knew its way around before it faced the question, 'how do I walk?', but in that reflective way that means that no intellectual holds are barred.[6]

Sellars explains that it "is therefore 'the eye on the whole' which distinguishes the philosophical enterprise" (p. 3). How does the philosopher keep his "eye on the whole"? One way is to do what Sellars proceeds to do in "Philosophy and the Scientific Image of Man" (the published version of the two famous lectures): explore the tension between what Sellars calls "the manifest image of man-in-the-world" and "the scientific image of man-in-the-world." What are these two images? They are two ideal types extracted for analysis from the history of conscious thought. They matter because they play a pivotal role in how we self-conceive. How we self-conceive matters greatly to how life seems or feels first-personally. Thus, how these images hang together or fail to hang together matters to how we fare subjectively. Although they bleed into one another, we can distinguish analytically among three images: the *original*, the *manifest*, and the *scientific*.

Imagine that when the ice melted at the end of the Pleistocene our cognitive schemes were only rich enough to enable us to achieve biological fitness. We knew where and when to forage and hunt. We made tools for these tasks and shared skills. And we possessed cognitive schemes that expressed the relevant know-how, perhaps not in a consciously expressible form. Our language and our expressive skills, we can imagine, were relatively immature and were devoted primarily to fitness-maintaining tasks. "Who are we?" and "What are we like deep down inside?" and "How are we situated in the cosmos?" are questions that may not have been asked or theorized. But whatever rule-governed ways we had then for getting around, for interacting with conspecifics, and so on, constituted the inchoate "original image" of ourselves and our world.

This original image develops and becomes more complex (which is different from "becomes more truthful") as we become more articulate in conceiving of our nature and our place in the cosmos. This sophistication of the original image in collective memory and narrative is shaped and conveyed in art, epic, fables, poetry, music, and spiritual practices. This is the *manifest image*. It is a work in progress, but one can catch it and examine it for a people at a time. When we talk about "how people see things," we are normally talking about the manifest image.

At some early stage in the development of the manifest image, what we now call "scientific thinking" is added to the mix. When this happens, either the manifest image absorbs science (as in the case of medicine and human anatomy) or the defenders of the manifest image try to smash what they perceive as threats to how human being is to be imagined (as in the cases of Galileo and Darwin). What science gets to say about human being is legislated by defenders of the manifest image.

Because science, with the imprimatur of the defenders of the manifest image, is absorbed into it, the manifest image cannot be said to be "unscientific." But a time comes when the scientific image develops autonomy from the manifest image, as well as a high degree of independent authority. Then there is serious competition between the images. Or so it seems.

Because our identity as humans is tied in essential ways to how we self-conceive, who and what we are seems confusing or bewildering, and our sense of our selves and our place in the universe seems unstable or (what is different) fractured. In a situation where there are two incompatible images on offer, dissonance can be resolved by placing all one's bets on one image over the other. Another tactic is to work to adjust both images so that they need not be perceived or experienced as inconsistent.

"The philosopher," Sellars writes (p. 4), "is confronted not by one complex many-dimensional picture, the unity of which, such as it is, he must come to appreciate; but by *two* pictures of essentially the same order of complexity, each of which purports to be a complete picture of man-in-the-world." The situation is that the two images are now equally authoritative but also not obviously mutually comfortable, consistent, or commensurable. Still, Sellars thinks the philosopher's job is to explain how these different images can both be true. Maybe.

Following the great physicist Arthur Eddington, Sellars compares the situation with two descriptions of a table: the solid table of common sense and the table made up mostly of empty space as conceived from the point of view of quantum physics. Could both pictures of the table be held in one's mind at once, or must we go back and forth between them as in a Gestalt illusion, alternating between the two images and never able to combine them in our minds at once? Is one picture what Nelson Goodman (1978) calls "the right version" of the way the world is,[7] or can multiple versions be true and useful for different purposes?

How Many "Worlds" or "Images" Are There?

I aim to examine the relation between the scientific image of persons and images that are available in other locations in our worlds. Specifically, my focus is on how contemporary mind science, informed by Darwin's theory of evolution, a sub-species of science conceived generally, interacts with some of the main locations to which we go to make sense of things and find meaning. Is neo-Darwinian mind science (which includes, but is not exhausted by, evolutionary psychology) a source of disharmony? If so, why? Are there ways to make the relations among what I call *spaces of meaning* more harmonious? Or is it all right if we live among and interact with disharmonious spaces? Does science generally, and do the human sciences specifically, disenchant the world?[8] To address these questions, I find it necessary to expand the dialectical space in which conflict, consistency, or consilience might be discovered or sought from a dyad to a sextet: {art, science, technology, ethics, politics, spirituality}. Each of these six spaces of meaning names, or gestures in the direction of, a large domain of life. Art includes painting, poetry, literature, music, and popular culture. Science includes all the sciences, as well as whatever synthetic philosophical picture of persons (or reality) is thought to emerge from the sciences. Politics includes the relevant local and/or nation-state form of government as well as the legal and economic structures it rests on and/or engenders. Spirituality includes multifarious religious practices and institutions, theologies, and such non-theistic spiritual conceptions as ethical naturalism, secular humanism, pagan shamanism, Confucianism, Buddhism, and Stoicism. The basic idea is that in order to understand how any group or individual self-conceives, what their practices of self-location, self-understanding, and their ideals for human development are, and how they work, one must give concrete values to these six variables. This will result in something like what Clifford Geertz (1973) called "thick description" (a phrase he took from Gilbert Ryle). Geertz's idea of thick description is connected to a wise observation made especially among anthropologists about their practice: Before one is able to say anything interesting about some individual's or group's thinking and behavior, one needs to understand what the individual or group is doing. This requires a rich and intricate understanding of the individual or group. One must understand behavior in terms of the pratices, traditions, and worldview that give the behavior sense and meaning. Thick description (*Verstehen*) involves a kind of understanding, albeit not necessarily a causal understanding, of some phenomena. When I tell you who I am, where I am coming from, how I think about things, and so on, I am providing a thick description. When I say what made me the way I am, I turn to causal explanation. Both are illuminating. The same principles apply to the six spaces of meaning. We live among them, and we will understand ourselves better if we understand deeply what each space affords and how it interacts with other spaces.

There are three main reasons for expanding the dialectical space from two spaces of meaning to at least six:

(1) The dominant form of dyadic analysis is to examine the conflict between science and religion. Indeed, there is a whole publishing industry devoted to the conflicts between science and religion. But the places in which people find dis-ease among the spaces of meaning involve more than just science and religion. I know artists who are not impressed by traditional religion (they are atheists) but who also find what they take to be the scientific picture incomplete or deflating. Familiarly, we say that the scientific description of the sunset, the tides, or consciousness fails to capture the phenomenon. Williams James believed that psychology as a science sensibly assumes determinism, "and no one can find fault," but that this deterministic assumption is not compatible with certain equally necessary assumptions about free will and moral responsibility that ethics makes. This conflict, which tortured James, is between science (specifically mind science) and ethics, not between psychology and religion. And, of course, there are abundant conflicts within and between nation-states that turn on conflicts between politics and religion, with science offstage, a nonfactor.

(2) When we humans conceptualize who we are and how we are doing, we do so in terms of narrative structures that have their homes in more than just the two expansive spaces named by religion and science. There is much recent interesting work in anthropology about the ways in which television engenders and reinforces everything from one's self-conception as a citizen to appropriate gender roles. (See Abu-Lughod 2001.) Whether television is intentionally used by nation-states for this purposes is a different and variable matter. The point is that television used in this way is doing more than, possibly something different than, articulating two ideal philosophical pictures of man in the world. It might be doing that to some extent, but as importantly it is depicting something more normative. It is recommending how one ought to be, live, and self-conceive as a man or a woman in a particular political community.

(3) Relatedly, the historical record indicates a persistent human effort to locate excellent ways of being and living. Each age seeks and articulates norms and ideals that if embodied would represent a good life, a life well lived, a beautiful and honest life. Speaking platonically,[9] we humans show persistent signs of relishing the adventure of trying to track down what is *good*, what is *true*, and what is *beautiful*. My six spaces of meaning connect up with these three forms in telling ways. Art tracks beauty—at least that is one of its functions. Science worships the true. Technology is useful, and what is useful is good in one sense. Ethics and politics track what is good in other senses of the term. And spirituality can be seen as a location in which all three forms are represented, although familiarly, in theistic forms. The spiritual space of meaning is charged by critics with caring little about what is true.

The overall warrant for beginning inquiry with six spaces or zones of meaning is that although most of us live our lives with our feet firmly planted on the ground worried primarily about friends, family, work, making ends meet and so on, how we live in ordinary life and how we experience ordinary life are affected by the multifarious ways we interact with these six spaces.

The claim that our lives are as they are in some measure because of the ways we interact with the six spaces of meaning is weak, first pass. I am not claiming that most people consciously analyze or score their lives in terms of how they relate to these (and only these) six spaces of meaning. We philosophers, with our heads in the clouds, might spend more conscious time than most folk among these spaces, wondering how they hang together (if they do hang together). One reason is that it is an occupational hazard that comes from thinking, following Sellars, that this is our vocation.[10] Philosophers aside, the claim is that the shape of perfectly ordinary lives is affected typically by commerce with the latter six spaces. My six spaces of meaning, like Sellars's two images, are inherently social. They are publicly available and we humans live among them. That is, we experience the world in and through these spaces. Regardless of whether a space or an image gets things right, we utilize modes it affords and/or recommends for self-conceiving.

A convincing example of how our narratives are affected (perhaps unconsciously) by ways of speaking and thinking that emerge from abstract spaces of meaning can be found in *Lolita*, where Vladimir Nabokov expresses the idea in a particularly vivid way. Humbert Humbert, reflecting on the narrative demands of his relationship with his wife, says:

> She showed a fierce insatiable curiosity for my past. She made me tell her about my marriage to Valeria, who was of course a scream; but I also had to invent, or to pad atrociously, a long series of mistresses for Charlotte's morbid delectation. To keep her happy, I had to present her with an illustrated catalogue of them, all nicely differentiated.... So I presented my women, and had them smile and sway—the languorous blond, the fiery brunette, the sensual copperhead—as if on parade in a bordello. The more popular and platitudinous I made them, the more Mrs. Humbert was pleased with the show.... Never in my life had I confessed so much or received so many confessions. The sincerity and artlessness with which she discussed her "love-life," from first necking to connubial catch-as-catch-can, were, ethically, in striking contrast with my glib compositions, but technically the two sets were congeneric since both were affected by the same stuff (soap operas, psychoanalysis and cheap novelettes) upon which I drew for my characters and she for her mode of expression. (Nabokov 1955, pp. 79–80)

This sort of use of "soap operas, psychoanalysis and cheap novelettes" is possible and natural because both Humbert and his wife live in a world that drips "Hollywood" and in which these modes of thinking and speaking play a prominent and possibly unseen (for Mrs. Humbert) role.

The passage from *Lolita* lends support to my idea that it will help in understanding how we humans make meaning and sense of things to expand the spaces utilized to ones like art (in this case soap operas and cheap novelettes) as well as science and technology (in this case the theory and practice of nineteenth-century Viennese psychiatry as adapted by Hollywood).

The members of the sextet, individually and collectively, name familiar domains with which virtually every modern life intersects. They name places we go to make meaning and sense of things, including ourselves.[11] Each space contains information about possibilities for self-description as well as norms for self-direction.

The Space of Meaning$^{\text{Early 21st century}}$

I call the sextet {art, science, technology, ethics, politics, spirituality} a *Goodman set* in honor of *Ways of Worldmaking* (1978), in which Nelson Goodman shows how such abstract social objects as spaces of meaning contribute to the constitution of our worlds. A world, or the multiplicity of worlds, in the relevant sense, is not the world in the sense of Earth, but the way 6 billion individuals live and conceive of their lives on Earth. A Goodman set is useful for locating—initially at an abstract level—the most salient spaces of meaning that make up the Space of Meaning for some social group. Correctly specifying the main spaces of meaning that some group uses to make meaning and sense of things points us ideally in the direction of the right socially constrained ways some group or groups of people conceive their world(s). Only if using some such analytic apparatus is legitimate for demarcating worlds can we say there are fewer than 6 billion worlds and give some conceptual traction to understanding our kind of being in the world.[12]

The members of a Goodman set can be individuated in a principled way by family-resemblance criteria, but not in terms of necessary and sufficient conditions. Each member names an abstract scheme, a "form of life," or an aspect of a form of life which humans intersect with, participate in, utilize, and deploy in making sense and meaning of things, including themselves. Each member is a variable with (to adopt a Ciceronian turn of phrase) the "customs and the times" giving values to the variables. This characterization of what a Goodman set consists of (spaces of meaning) and what it taken as a whole is (the Space of Meaning for a group) has clear empirical implications. First, people should generally know how to classify items under the right general category in forced-choice tests. But there will be indeterminate cases, or, better, cases that fit in more than one category—for example, many of Leonardo Da Vinci's drawings are sensibly classified as art and science. Second, people will speak in ways that reflect which spaces are getting attention or having an impact (perhaps short-lived). In the 1980s the vocabulary of input-output ("Thanks for the input") from computer science replaced the 1970s stimulus-response vocabulary from behaviorism, which had replaced the more aesthetically and politically inspired 1950s (beatnik) and 1960s (hippie) ways of speaking. Third, the spaces, as well as what falls under them, will differ for different times and places. Science in 2007 includes neuroscience and genetics; it included neither a century ago. The Goodman set that constituted the Space of Meaning for twelfth-century Europeans would not include science at all. Art now includes a musical form called "rap" or (what is somewhat different) "hip-hop." It didn't in 1987. Monotheism is a relatively recent spiritual option, only 2,500 years old. And so on.[13]

The central claim is that it is, to some significant degree, by living in these spaces that we make sense of things, orient our lives, find our way, and live meaningfully. Each member of a Goodman set is a space of meaning. A Goodman set of spaces of meaning correctly characterized for some group is the Space of Meaning for that group. The Goodman set above of {art, science, technology, ethics, politics, spirituality} constitutes what I call the Space of Meaning$^{\text{Early 21st century}}$. Most contemporary people interact with all six of these spaces.

I say 'most' rather than 'all' because a substantial number of Earth's inhabitants—perhaps 20 percent—do not interact in a full or rich way with the Space of Meaning$^{\text{Early 21st century}}$. These are fellow humans who live in a condition of "absolute poverty" as defined by economists. Insofar as they can be said to live in a Space of Meaning at all, it is probably best conceived as dominated or constituted by a spiritual or religious view that provides some small (albeit possibly false) hope against their otherwise utterly hopeless lives.[14] Such lives are objectively awful, although if we imagine (as we should) how to help such souls to escape absolute poverty we might legitimately wonder whether it would be a good thing for such people to eventually interact with all the spaces of meaning that the average American does. Besides the problem of living meaningfully for those who live in conditions of absolute poverty, there are also serious problems in racist or sexist nations (virtually all nation-states) where, even if there is no absolute poverty, there are discriminatory practices that keep certain groups worse off than others in wealth, and in addition give these groups less voice in creating, modifying, and participating actively in the Space of Meaning$^{\text{Early 21st century}}$ as it is embodied in their homeland. Members of such groups are "spiritually" worse off than their compatriots in virtue of social practices that circumscribe how they are permitted to interact with the Space of Meaning$^{\text{Early 21st century}}$. I will return several times to this question of our responsibilities to those who live in conditions of material or spiritual poverty.

Meaning Pluralism and Meaningful Relations

One additional reason for broadening the scope of inquiry from a conflict between two images or spaces of meaning, science and religion, to six (or more) deserves emphasis, since from this point forward I will largely assume it. The reason has to do with a commitment to the idea that there are plural ways of making sense of things and finding meaning. This is because there are in reality a multiplicity of kinds of things (kings and cabbages and numbers) and relations. Different spaces are suited to speak most profitably about different relations. One reason one ought to be a space-of-meaning pluralist has to do with the ontology of relations.

Science specializes in the causal relation. Some who fear that the scientific image is reductive or eliminativist, are worried about the tendency of certain scientistic types to think that the only real relation, or the only interesting one, is the causal relation.

Here is the right reply: Even if everything that there is is the way it is because some set of causes made it that way, it does not follow that the only real relation or the only interesting relation is the causal one. Science itself recognizes numerical, spatial, and temporal relations that are not causal. Atom a is *next to/closest to* b. There are *eight* distinct atoms left in the chamber. Atom a moved *after* b hit it.

Because some very important relations are causal, and because science is especially good at uncovering causal relations, science is very important to understanding things.[15] But there are many other types of relations than causal ones. There are arithmetic, geometrical, and logical relations (e.g., if p then q, p, therefore q). There are statistical relations, aesthetic relations, personal relations, semantic (meaning and reference), syntactic or grammatical relations, ethical relations (action a is better than action b), and so on.

There is nothing spooky about there being more relations that are real, and that matter, than relations that are causal. Furthermore, we are good at tracking all the latter relations, and doing so helps us to make sense of things and find meaning.

The Psycho-Poetics of Experience

"The central claim," I wrote above, "is that it is, to some significant degree, by living in these spaces [of meaning] that we make sense of things, orient our lives, find our way, and live meaningfully." The psycho-social picture is this: We humans are creatures who live as beings in time with our feet on the ground, interacting in and with the natural, social, and built worlds. Living is a psycho-poetic performance, a drama that is our own, but that is made possible by our individual intersection, and that of our fellow performers, with the relevant Space of Meaning. For us contemporaries, how we act, feel, move, speak, and think in the world depends in some measure on how we weave a tapestry of sense and meaning by participation in various subspaces within the spaces of meaning that constitute the Space of Meaning$^{\text{Early 21st century}}$. Did people always conceive of life artfully? I don't know. It doesn't matter. We now do. This is why it matters, why it would be good, if we could gain some clarity on this question: How, if they do hang together, do such non-thingy things as the practices, forms of life, and ways of world-making that shape and partly constitute our individual psycho-poetical performances interact, intersect, and hang together? These non-thingy-things are the stuff of schemas, cognitive models, forms of life, world hypotheses, modes of inquiry, disciplines, *Weltanschauungen*, the Background, the Horizon, social imaginaries, master narratives, and meta-narratives. They form at least a significant part of the Background within which we live our lives. But they are all "on the move." So they are also a Foreground, places we extend our selves into—the Horizonal Zone. These non-thingy things all have visible public lives, at least in the West, and, I think, in all three other geographical directions as well.[16] How seriously and respectfully each is treated, how much each aspiration or set of practices is socially supported, is, however, a matter of considerable variability. And, again, how and to what degree any particular individual "participates" in these spaces, or creates his or her own psycho-poetic performance by intersecting with them, is variable. The variability is one way we express our individuality. The main point is that how my life goes depends in some measure on how I self-conceive. How I self-conceive depends in some measure on the spaces of meaning.

The *Lebenswelt*

Edmund Husserl called the individual instantiation of life among the spaces constituting a Space of Meaning the *Lebenswelt*. For each person in, say, a particular community, there is a *Lebenswelt* $(l_1, l_2, \ldots, l_n)$ that constitutes the lived world, the psycho-poetic performance for each individual. The lived world has a subjective, something-it-is like nature, which is the way it is experienced first-personally, as well as an objective side, which is captured by the individual's enactive, embodied being in time in the world. Some of the things we do, we know about and understand first-personally. Sometimes third parties understand us better than we ourselves do.

The collective Background, as well as the Horizon or set of horizons they gesture toward, constitute a vast public space—a space that no individual could possibly comprehend in full. Furthermore, in part because it is vast and social, the Background is not always well articulated, and it includes a certain amount of meta-theory. There is art, but there are also theories about what art is and isn't, views about kinds and degrees of beauty. Aesthetic theories. There is science and there is philosophizing about science. There are abundant technologies and there are widespread, taken-for-granted assumptions about what sorts of technologies are absolute necessities—for example, televisions and telephones. There are the actual effects that living with the Internet has on lives, and there are academic conferences that theorize and articulate these effects.

The Background, as a container of theory(ies) and meta-theory(ies), has its origins and roots in embodied human practices—in the production of art, in scientific theorizing and experiment, in utilizing technologies for work and entertainment, in moral education, in political debates, in the enactment of legislation, in spiritual experience, meditation, and prayer, and in the building of sacred spaces. The spaces of meaning are created collectively and emerge from collective activity. They then, as emergent products, grow and develop and constitute spaces we each enter to make a life, to live out the psycho-poetic performance that is our life.

The Space of Meaning$^{\text{Early 21st century}}$ is abstract and intentionally so. Indeed, it is its abstract quality that makes it useful and allows descent to more grounded places such as the lived worlds of individuals. Let me explain.

'Psycho-poetics' refers to the creative ways persons attempt to make meaning and sense of things and thereby to live well. A person who lives well, in a way that makes sense and is meaningful, is what the Greeks called 'eudaimon'—literally, "happily blessed." Eudaimonia is flourishing. Aristotle said, and I agree, that all humans seek eudaimonia, although importantly they disagree about what makes for eudaimonia. If there can be such a thing as eudaimonics, systematic theorizing about the nature, causes, and constituents of human flourishing, it is because it is possible to say some contentful things about the ways of being and living that are likely to bring happiness, sense, and meaning to persons.

The compound term 'psycho-poetics' is designed to draw attention to the fact that the human attempt to make meaning and sense of things is akin to a performance executed ideally with style, grace, feeling, and a certain amount of mindfulness. To say that persons are engaged in psycho-poetics has a descriptive and a normative component. Individuals co-create their performance with others inside the space of socially available modes of being, thinking, and feeling. Furthermore, this is something we ought to do, mindfully at times, in order to maximize chances of living meaningfully and flourishing.

Life among the Spaces

Ordinary lives necessarily engage three worlds: a natural world, a built world, and a social world. There are the very concrete activities of eating, drinking, making love, making babies, making a living, working, engaging in hobbies, being friends, being enemies, and burying loved ones. Living life on the ground involves doing these things. Emphasizing this might make one press this concern: What does the Goodman set that constitutes the Space of Meaning$^{\text{Early 21st century}}$ have to do with actually living an ordinary life as most people live such lives? The answer is this: In living our lives, and in speaking with others about our lives and theirs, we take our ways of speaking and thinking, as well as our norms, to some significant degree from the relevant spaces of meaning that constitute our Space of Meaning.[17]

Of course, in relation to some of the six spaces that constitute the Space of Meaning$^{\text{Early 21st century}}$ we are actively engaged, whereas in relation to others we are audience, sometimes inattentively or disinterestedly so. Among the 80 percent of Earth's population who have enough to survive, few devote equal time and energy to each member of the Goodman set that constitutes the Space of Meaning$^{\text{Early 21st century}}$. One reason is that no one deems each equally significant to finding his way. Especially in the West, where no one needs to live in absolute poverty and where communication media are intrusive and speak about all these forms of life and all these ways of being, it is hard to be completely oblivious even to domains one cares little about. Thus, it is a rare bird who does not intersect and interact with most of these six social spaces in some way or other. Such interaction, such intersection, is so much expected that the norms governing our ideals of good and meaningful lives require that we interact in some way or another with most of these spaces and that we be able to narratively track to some degree how we are doing so. The self-expressive, self-locating narrative by which we describe who we are, where we come from, and where we are headed is by and large the report on our own psycho-poetic performance. (On the connection between narrative and selfhood, see Dennett 1988, 1991; Flanagan 1991a,b, 1992, 1996a, 2000b, 2002; Fireman, McVay, and Flanagan 2002.) Indeed, each individual is in some significant way the person constituted by this psycho-poetic performance.

Starting at the abstract level, thanks to the inclusiveness of the superordinate categories, allows us to think, speak, and compare what the psycho-poetic performances are like for individuals by descending from the abstract to where each lives among the spaces in the Space of Meaning$^{\text{Early 21st century}}$.

Consider the following three lives as a way of seeing how moving between the concrete *Lebenswelten*, the psycho-poetic performances of individuals and cohesive social groups, and the abstract Goodman set that constitutes the Space of Meaning$^{\text{Early 21st century}}$ provides the right sort of analytic space for the present inquiry. A male Celtic-Catholic-Buddhist from Durham who has raised two atheists of great charm and integrity, who does philosophy, is impressed by and knowledgeable about biology and mind science, loves both Bach and the Beatles, and is on the political left participates in the Space of Meaning$^{\text{Early 21st century}}$ in one recognizable way. A female Muslim from Dearborn who works as an engineer, enjoyed *Reading Lolita in Tehran*, paints in watercolor, has a son in the U.S. Army, supports the war in Iraq, and has raised her children to be devout participates in the Space of Meaning$^{\text{Early 21st century}}$ in another recognizable way. We can live happily in the same country, making meaning and sense of things. Could we marry and live happily ever after? Doubtful. A Maasai thirty-something who runs safaris and Kilimanjaro ascents from Dar es Salaam, listens to African hip-hop and West Indian reggae, sculpts Maasai folk images in teak, has deep knowledge of the flora and fauna of Tanzania and Kenya, and works at a distance against the genocide in Sudan can also be easily seen as working in the Space of Meaning$^{\text{Early 21st century}}$.

Despite the fact that the six spaces of meaning that constitute the Space of Meaning$^{\text{Early 21st century}}$ (I don't claim that the list is exhaustive) are all abstractions, they are useful abstractions. In the language of linguistics, the name for each space—'art', 'ethics', 'science', and so on—is a superordinate term, as are 'vehicle' and 'job'. Cars, trains, airplanes, jets, motorcycles, and rickshaws are all vehicles. A 2004 Vespa Serie Americana motor scooter is my vehicle. Fireman, policeman, carpenter, stockbroker, farmer, doctor, and lawyer are all jobs. None of those is my job, which is teaching and doing research at Duke University.

Starting with the superordinate categories, even though it entails that we are starting the conversation in abstract space, has several advantages. First, public discourse about conflict between spaces commonly occurs using exactly these abstract terms. Consider, for example, the alleged conflict between science and religion. There really isn't any such conflict, since neither science nor religion names a single, determinate, or homogeneous practice. There is, as I write, a conflict between Darwinism and creationism and intelligent design, especially in the United States. But chemistry, anatomy, and medicine are parts of science, and they are not bothering most religious folk. Second, not all spiritual traditions are having trouble, or need to have trouble, with evolution. If certain Christians stopped claiming that the Genesis story (which of the two?) is literally true, part of the problem would disappear. The Dalai Lama is pretty comfortable with evolution. Many spiritual folk, the Earth over, have not yet heard of or absorbed the theory. Time will tell how they respond. So one advantage of starting at the high level is that when conflict occurs, the abstract taxonomy composed of the six spaces of meaning allows us easy descent to the exact location of the problem. Some philosophers, as well as some literary and art theorists, speak of the conflict between art and (conventional) morality. But in almost every case I can think of we need to descend from the superordinate spaces named 'art' and 'ethics' and get into nitty-gritty discussions of, say, Plato's objections to poetry as it pertains to the moral (mis-)education of the youth, or of rap and hip-hop music as pernicious reinforcers of sexist or homophobic beliefs and practices, or of whether Andres Serrano's "Piss Christ" is disrespectful or sacrilegious. If after discussion the problem seems bigger, such as the sense that science in general is disenchanting or that art generally disrupts ethics or politics or undermines religion, then we can ascend and talk about that.

A final point relating to the interplay between the abstract and the concrete: One might think I would be wisest to narrow the topic to the highly visible conflict between science and religion because that is the region in which the most contentious debates about human flourishing and life's meaning seem to occur. However, this visible and noisy conflict may not be as deep or widespread as it seems. Truth be told, I fear that if I give it more attention than it deserves, I encourage the conflict in just the way one encourages an occasionally naughty boy by giving him too much attention when he is naughty. Second, I am certain that we will do best if we frame whatever conflicts exist between this dyad within the wider space of the multifarious things we do to make meaning and sense.[18] Forget for a moment about science and its relations to religion. If one conceives of science as the only epistemically "first-class" way of speaking (I believe Quine used this expression for physics and he had no hope that any human science could ever achieve "first-class" status), it is not at all clear how ethics and politics are to be conceived. Ethics and politics have to do with virtues, values, norms, and practices that are productive of the common good. They ask the perennial questions: How shall I live? How shall we live? Ethics and politics had better be cognitively respectable if eudaimonics is possible.

Many scientists will claim that science is unopinionated on virtue and vice, human flourishing, and the like. But if we grant to science the broad scope that global metaphysical naturalism and scientism seem to entail, then it is not clear how anyone could be legitimately opinionated on such matters. Global metaphysical naturalism is an imperialistic ontological view of maximal scope: What there is, and all there is, is matter and energy transformations among natural stuff. Scientism says that everything worth expressing can be expressed in a scientific idiom. If either of these views is credible, or if both of them are, it is hard to see what ethics and politics are, do, or are about.

Similarly for the arts. Music, literature, poetry, painting, drama (serious and comic), and dance are all ways in which and through which humans try to make meaning and sense. (This is so whether one is situated as artist or as audience.) What is art? What is it for? Picasso's *Guernica* or Munch's *The Scream* is said to speak truthfully about something. How can art speak truthfully about war and existential despair if everything real is no more than matter and energy transfers among natural stuff? What could war and existential despair even be? If the arts are speaking about matter and energy transfers, that they are doing so is well disguised, and we are seriously confused about what they are doing. Politics (and I guess ethics too) could be analyzed as forms of engineering, something science makes sense of, indeed that it gives rise to. Machiavelli, before science was really big, had this idea. And Quine advanced the idea late in the last century. But if we allow that politics often expresses, and occasionally embodies, views about goodness and beauty, and is not exclusively concerned with matters of social coordination, conflict management, and the like, then the problem of the place of these things in the world that science purports to describe and explain resurfaces.[19]

Geisteswissenschaften: Our Peculiar Situation

The six ways of making meaning and sense that constitute our Goodman set and thus the Space of meaning$^{\text{Early 21st century}}$ all have long histories. But perhaps only in the West has science been on the list for several centuries. *Naturwissenschaften* blossomed in the seventeenth century. The nineteenth century marked the official appearance of *Geisteswissenschaften*, anthropology, sociology, and psychology as well as new ways of conceiving of history and political science as scientific or potentially so. *Geisteswissenschaften* joined *Naturwissenschaften* in the pursuit of describing and explaining (and in some cases predicting and controlling) whatever can be described and explained naturally. This fact created a special situation. In our time, the human sciences—especially but not exclusively the mind sciences—are opinionated about the nature and status of the other ways of making meaning and sense. Indeed, the very idea of the human sciences implies that all human practices can, in principle, be understood scientifically. Here is the Possibility Proof:

1. Humans are natural creatures who live in the natural world.
2. According to the neo-Darwinian consensus, humans are animals: *Homo sapiens sapiens*, mammals who know and know that they know.
3. Human practices are natural phenomena.
4. Art, science, ethics, religion, and politics are human practices.
5. The natural sciences and the human sciences can, in principle, describe and explain human nature and human practices.
6. Therefore, the sciences can explain, in principle, the nature and the function of art, science, ethics, religion, and politics.

Explaining Ways of Worldmaking

What might explaining our practices—our ways of worldmaking—involve? Presumably we would try to understand the nature and functions of these practices, as well as their causal antecedents and consequences. This would lead us to understand the nature of *Homo sapiens* more deeply. It would almost inevitably require changes in traditional narratives of self-understanding. If the changes involve filling in blank spaces, that is good. Knowledge is increased. If, however, well-entrenched views about the nature of our world and ourselves are asked to yield to better ways of understanding, the task is more complicated and stress-inducing. It may involve revising stories that we think of as necessary for living meaningfully.

One surprisingly common idea is that science, in explaining some phenomenon, makes it something it isn't or wasn't. It tries to disclose that every thing is a "mere thing." It takes the world as we know it and turns it into a mere collection of scientific objects. 'Reductionism' is the disparaging name for this phenomenon. Something like this view—that reduction always entails that things are not as they seem, and that such phenomena as consciousness are revealed as illusory—is common. But it rests on a mistake. To say that some phenomenon can be understood scientifically, even that it can be reduced, is not to say that the phenomenon is itself "scientific," nor does it entail that the phenomenon we began with disappears or evaporates—whatever exactly that might mean—when we get at its deep structure. Consider a simple case: Water is H_2O. Water is not explained away; its nature is understood more deeply. Water is a natural element. It is the explanandum. H_2O is the explanans. Is either water or H_2O itself "scientific"? The question makes no sense. Water is a natural phenomenon, and science helps us to understand its microstructure, which explains why it in fact possesses such higher-level properties as fluidity. That's all there is to it.

The Threat of Scientism

Scientism is the source of some of the dis-ease with contemporary science. Scientism is the brash and overreaching doctrine that everything worth saying or expressing can be said or expressed in a scientific idiom. It is arguable that some of the European logical positivists of the 1920s and the 1930s came close to embracing scientism.

The claim that science can, in principle, explain everything we think, say, and do—that it can, in principle, provide a causal account of human being (a causal account of *Dasein*)—should be distinguished from the claim that everything can be expressed scientifically. Consider art and music. It is patently crazy to say that the works of Michelangelo, Da Vinci, Van Gogh, Cezanne, Picasso, Mozart, Chopin, Schönberg, Ellington, Coltrane, Dylan, or Nirvana could be expressed scientifically. Assuming something like the best-case scenario for science, we might want to say that artistic and musical productions can be analyzed in terms of their physical manifestations—painting in terms of chemistry and geometry, and music in terms of sound waves and mathematical relationships.

Furthermore, some very complex combination of the culture, individual life, and the brain of some artist might allow for something like an explanation sketch of why that artist produced the works he or she did. Kay Redfield Jamison (1993) has done very interesting work on the high incidence of bipolar disorder among great nineteenth- and twentieth-century poets and musicians.[20] Such work might lead us to understand more deeply what ordinary and creative imagination consist in. But such work does not replace or reveal what Walt Whitman, T. S. Eliot, W. B. Yeats, Dylan Thomas, Sylvia Plath, or Seamus Heaney says, means, or does in the language of poetry.

There is nothing remotely odd about these kinds of scientific investigation of art or music, or of the creative process itself. But although such inquiry takes artistic or musical production as something to be explained, it does not take the production itself as expressing something that can be stated scientifically. The claim that not everything can be expressed scientifically is not a claim that art, music, poetry, literature, and religious experiences cannot in principle be accounted for scientifically, or that these productions involve magical or mysterious powers. Whatever they express, it is something perfectly human, but the appropriate idiom of expression is not a scientific one. The scientific idiom requires words and, often, mathematical formulas. Painting, sculpture, and music require neither. Indeed, they cannot in principle express what they express in words or mathematical formulas. Therefore, whatever they express is not expressible scientifically. To be sure, poetry, literature, and music with lyrics use words. But their idiom is not a scientific one. And the reason is doubly principled: Many of the relations explored are not explored causally (the relation in which science excels). A good love song can make you feel love, but it never does so by getting into the "pheromonics" and the neurobiology of love. The arts work our imaginations with all the playful tricks of language, allegory, metaphor, and metonymy that science, for it purposes, doesn't much care for.

Historians of literature and art often tell us useful things about art—for instance, about how poets and artists were influenced by scientific ideas—and psychologists can explain important things about the physiology of perception. Despite the illumination provided, neither provides anything approaching a complete or satisfying explanation of what any interesting artistic work means or does. The simple and obvious point is that not everything worth expressing can or should be expressed scientifically. Scientism is descriptively false and normatively false. This, I like to think, will quell some of the anxiety. I like this outcome because temperamentally

I don't like for people to be anxious. If there were a basis for legitimate fear and trembling, sickness unto death, and the like, I would have to say so. But there isn't, so I don't.[21]

The Scientific Image of Persons and Big Mistakes

The bugbear of scientism aside, what about the picture of persons that Sellars calls "the scientific image of man-in-the world"? Does the scientific image reveal any deep or big mistakes in our ordinary folk-philosophical picture of persons? Is the scientific image disturbing, demeaning, and disenchanting?[22] These last two questions are logically distinct, but in fact they connect up. If science, or the image it projects, said that there are no persons, or that we *seem* conscious but aren't, or that we can never act freely, then these are "truths" that *if true* would be disturbing.[23]

What does contemporary mind science say about the nature and function of consciousness and about mental causation, about human agency? Does deeper scientific understanding of consciousness and mental causation help or hinder our efforts to make sense of things, to find meaning, and to flourish?

Sellars, recall, says that "the philosopher is confronted not by one picture . . . but by *two* pictures of essentially the same order of complexity, each of which purports to be a complete picture of man-in-the-world, and which, after separate scrutiny, he must fuse into one vision" (p. 4). In view of the possible conflict between the manifest and scientific images on the issues of consciousness and causation, one might wonder what the force of 'must' is here. If the scientific image insists with good reason that some essential tenet of the manifest image is false, then the two images cannot be fused.

Perhaps it is best to read Sellars's statement that the philosopher must "fuse [the two images] into one vision" as meaning that the philosopher ought to try to fuse the two images. But if the philosopher can't do that, something must yield.

Remember also that for Sellars there are actually three ideal types dubbed images. The "original image" was magical in the sense that essentially all moving objects, even abstract objects such as thunderstorms and seasons, were personified. "From this point of view," Sellers writes, "the refinement of the 'original' image into the manifest image, is the gradual 'depersonalization' of objects other than persons. That something like this has occurred with the advance of civilization is a familiar fact. Even persons, it is said (mistakenly, I believe), are being 'depersonalized' by the advance of the scientific point of view." (p. 10) This passage suggests that Sellars might think that the two images can be fused, and thus that the effort to do so can succeed because the scientific image does not "eliminate" the category of person. If it did, the two images would be inconsistent, since the manifest image treats the concept of person as fundamental and ineliminable. But even if neither the manifest image nor the scientific image denies that there are persons, there is this difference: The manifest image assumes "that what we ordinarily call persons are composites of a person proper and a body" and that "the essential dualism in the manifest image is not that between mind and body as substances, but between two radically different ways in which the human individual is related to the world" (p. 11).

It is interesting and instructive that Sellars, circa 1960, sees the manifest image this way, insofar as Gilbert Ryle, writing a decade earlier, read the manifest image as Cartesian to the core. Ryle called substance dualism "the official view" (and, less kindly, "the myth of the ghost in the machine"). Sellars acknowledges that philosophers read the manifest image through Cartesian lenses, but he judges the dominant common-sense position to be Lockean. Locke, recall, pleaded agnosticism (in some moods) on whether our two "radically different ways" of conceiving of humans—as continuous bodies and as continuous "persons"—depend on one substrate or two. I am not going to fuss over the question of what kind of dualism the manifest image incorporates, although I do think it is typically some form of immaterialism, either immaterial substances or immaterial properties.

Happily, the scientific image does not claim that there are not persons, but it does reject dualism. That is, it endorses (at any rate, this is the consensus position) some form of materialism or physicalism about persons. Is this the end of the world as we know it? I think not.

Subjective Realism, Neurophysicalism, and Phenomenal Consciousness

One thing many people fear about a naturalistic view of mind is that it will, in virtue of identifying mind with brain, make experiences a thing of the past. I have experienced at first hand being introduced as a "neurophilosopher" at receptions where people then proceed to treat me as an anthropological specimen who must have no inner life and who must believe that they are zombies! The worry, as best I can tell, goes something like this: A dualist picture of mind insists that we humans possess phenomenal consciousness. There is something it is like first-personally to be a subject of experience. We are not mere information processors. We have experiences. The scientific picture of mind identifies the mind with certain objective physical processes. But the subjective and the objective can't be meshed or melded. Not only isn't first-person phenomenal consciousness not "objective"; it cannot even, in principle, be captured in the sort of third-person objective description that normal science relishes.

The belief that consciousness can't be explained scientifically is also held by a surprisingly large number of naturalistically inclined thinkers. Thus, one hears much talk of the unbridgeable "explanatory gap" between subjective experience and the objective features of brains, and much about the "hard problem of consciousness." In 1991, I dubbed naturalists who think consciousness is a natural phenomenon that can't be explained scientifically "mysterians."[24] One kind way to describe them is as believing that there are epistemic limits, such as Heisenberg limits or Gödel limits or the limit that can't get us closer than 10^{-43} second to the Big Bang, that are relevant to our ability to solve the problem of consciousness. However, no mysterian has provided any Gödel-style proof that this is so. Instead, mysterians count on intuitions of the form "to say that my perception of blue is realized by activation of the blue-detecting neural network in my visual cortex is very unsatisfying." My response, then and now, is that intuitions like this should not be trusted, especially ones predictably hardened by long adherence to dualist views of mind, and furthermore that there is all the difference in the world between an explanation that is intuitively satisfying and one that is scientifically satisfactory.

No one, dualist, naturalist, or pan-psychic, has yet explained consciousness. What we have are pictures of how we might explain it, and differing assessments about how far along various research programs are in the attempt to explain how and why experiences occur. What we can say for certain now is that a naturalist picture fits much better with the rest of science than any known dualist or immaterialist view. Of course, the naturalist must accept the burden of showing how, using only natural resources, he proposes to explain phenomenal consciousness. Here is how.

Token neurophysicalism is the view that each and every mental event, each and every experience, is some physical event or other—presumably some central-nervous-system event.[25] *Type neurophysicalism* is the view that each kind or type of experience, e.g., "seeing a red cube" or "believing that [snow is white]," each kind or type of event—be it perceptual, emotional, or a belief—is realized in "pretty much the same way" by each member of the species that has the relevant experience. Type neurophysicalism appears to be true for rhesus macaques in narrow experiments where the task is to alternately detect (i.e., experience) vertical or horizontal lines—that is, distinctive populations of neurons fire in the same area depending on which sort of lines they see. For present purposes it does not matter whether "the same" mental state ("seeing red" or "believing that [snow is white]") are realized in different ways by human brains or are realized in very similar ways and/or locations in human brains. What matters is that each and every experience supervenes in some strong sense on a brain state. We can accept the truth of token neurophysicalism, and thus reject all immaterialist views that deny it, while resisting the conclusion that the essence of a mental event is revealed completely or captured completely by a description of its neural level realizer. The reason is as follows, and it applies uniquely to conscious mental events: Conscious mental events are essentially Janus-faced and uniquely so. They have first-person subjective feel *and* they are realized in objective states of affairs. As John Dewey said (1922, p. 62), "given that consciousness exists at all, there is no mystery in its being connected with what it is connected with."

Speaking counterfactually, water would be H_2O and gold would be the substance with atomic number 79 even if there were no subjects of experience, no sentient beings, in the world.[26] *Objective realism* is true of water and of gold.

But even if a conscious-mental-state token (say, your experience here and now of seeing these words on this page) is realized, and realized necessarily, in some complex neural process *n* in you, it is not the case, speaking counterfactually, that *n* could occur in a world without subjects. Specifically, *n* could not and would not occur in a world in which you were not reading these words. It is fine with the token physicalist if for each of us the neural realizer of the experience of reading the exact same sentence on this page, is somewhat different so long as there is some neural realizer or other that is the experience of seeing or reading that sentence for you.

The objective states of affairs in brains that *are* conscious mental events (not all, even widespread, neural activity is conscious) are unique in producing first-personal feel—*phenomenality*.[27] If certain objective states of affairs obtain, then so do first-person feels, and if there are first-person feels, then the relevant objective states of affairs obtain.

The asymmetry between water and gold, on the one side, and conscious mental events, on the other, can be said to come to this: the nature of water and gold is essentially objective—it is completely objective, ergo objective realism. The nature of conscious mental events is such that despite being perfectly natural, objective states of affairs, they have as part of their essential nature the subjective feel they have.

Call the basic idea *subjective realism*. Subjective realism says that the relevant objective state of affairs in a sentient creature properly hooked up to itself produces certain subjective feels in, for, and to that creature. The subjective feel is produced and realized in an organism in virtue of the relevant objective state of affairs' obtaining in that organism. The subjective feel is, as it were, no more than the relevant objective state of affairs obtaining in a creature that feels things. However, since the relevant objective state of affairs is only "captured" as the thing it is (in this case, a conscious mental event) as it is "captured" or "felt" by the organism itself. Thus, a completely third-personal neural description or causal explanation of an experience doesn't "capture" it as the experience it is. The reason is that third-personal descriptions don't "capture" feels. Certain third-personal states of affairs are the realizations of feels, but the feels are only had or captured by (or in) the creatures in whom those states of affairs obtain.

Suppose *β activity* is how seeing blue is realized for *Homo sapiens*. That is, suppose that, unlike in cases of semantically decoding spoken or written speech, the cell assemblies that underlie color perception are the same across members of our species.[28] We can then say that when Bert sees blue he is in the β state. The β state is how seeing blue is realized in all people. But Bert's seeing blue despite being realized by β activity in him is not realized solely in virtue of being β activity; it is realized in virtue of being β activity *in him*. And it is in virtue of being realized *in him* in the right way that he sees blue. Bert's seeing blue is nothing more than Bert's being in a certain objective psychobiological state. But it is a state that produces, or better, that has as an essential feature, a certain feel for Bert. How and why it does so is, I take it, explainable fully in naturalistic terms. Imagine that there is a complete neural description of what is going on in Bert—a complete description of β activity as it is uniquely realized in his nervous system. This description as offered from the third-person perspective completely captures the fact that Bert is seeing blue. Indeed, if the entire causal picture from the external blue object to his experiencing it were filled out we might claim to have explained fully why Bert is experiencing blue. But neither description "captures" what it is like for Bert to see blue. The experience is only captured by Bert first-personally. It is not important that Bert be able to say anything deep or interesting about what his experience is like. It is enough that he experiences blue or is seeing things bluely. This is sufficient, I hope, to see how one might be committed to the truth of *neurophysicalism* (token or type) about the conscious mind without being committed to the claim that the essence of an experience is captured fully as the experience it is by describing completely its neural realizer.[29]

For many it produces a mental cramp to think the thought that mental events are neural events but that their essence cannot be captured completely in neural terms. Such is the power of *objective realism*, a doctrine that is true for most of the things and types of things in the universe but that is not true for experiences. The cramping can be eased, I propose, by accepting that the subjective realist is claiming nothing mysterious. It is simply a unique but nonmysterious fact about conscious mental states that they essentially possess a phenomenal side. Don't mention that, and possibly how, they appear first-personally and you haven't described one, possibly two, of their essential features. Your metaphysic is incomplete. See things in the Janus way recommended and the intuition that gives rise to the thought that there is an unbridgeable explanatory gap between conscious mental states and their realizers is deflated, and it may disappear. Or so it seems to me.[30]

The main point of this part of the therapeutic exercise can be summed up this way: The individual gripped by the idea that we possess non-physical minds (or that mind involves non-physical properties) makes this sensible demand on any naturalistic view that could even be entertained as a replacement view: "Don't mess with phenomenal consciousness. It is a given that all my compatriots and I are subjects of experience. So you will need to say more than that a naturalistic neurophysicalist conception of mind is simpler than a dualist view to remotely capture my interest. There are all sorts of views that are simpler than their opponents—for example, that water is the only element is simpler than every view which countenances more than one element—but that fail because they are miserably simplistic. The simplicity of a view is only an interesting feature of that view when it explains everything that both views agree needs explaining. And in the case of mind, one thing, perhaps the main thing that needs explaining is how experience is possible, how there could be phenomenal consciousness in a material world." But we have now seen how this can be done. The subjective realist is a (neuro) physicalist who claims to be able to meet the plausible demand of the Cartesian, or any other variety of dualist, to provide a theory sketch or a plausibility proof, or at least to leave ample space for phenomenal consciousness. For the subjective realist, as for the immaterialist, it is a fundamental fact that phenomenal consciousness exists and is in need of explanation. My own view is that it is a law of nature that humans, and all other creatures that have experiences, have their own experiences in virtue of the way they are hooked up to themselves and to the world. *Homo sapiens* is the name for a class of creature that each has his or her own nervous system and no one else's. Thus, each person has his or her own experiences and no one else's. Something like subjective realism and something like neurophysicalism are the best candidates for the regulative assumption under which neuroscientists operate.

Mental Causation and Free Will

Like James (1890) and Kim (2005), I think there are two equally important mind-body problems: (1) the mind-brain problem, specifically what Dave Chalmers (1995, 1996) calls "the hard problem of consciousness," and (2) the "mental causation" problem. I have just sketched (yet again; see Flanagan 1991a, 1992, 1996a, 2000b, 2002) how I think the problem of consciousness can be solved naturalistically. My view leaves consciousness, at least how experience *seems*, exactly as before. The only thing it requires adjustment to for the ordinary person is his or her view on the constitution of consciousness, how conscious mental events are realized.

What now about mental causation? Here too I propose to naturalize the concept of free will—that is my job, after all (Flanagan 1991a,b, 1992,

1996a, 2002). And although I know doing so does not change *my* conception of what I am up to when I act freely, experience has taught that my way of speaking is unsettling and really does not seem to leave the phenomena and the associated practices intact. I'll let the reader decide.

Thomas Nagel concisely expresses one major worry that involves the question of whether a scientific view of human action could conceivably be fused with our ordinary view:

> If one cannot be responsible for consequences of one's acts due to factors beyond one's control, or for antecedents of one's acts that are properties of temperament not subject to one's will, or for the circumstances that pose one's moral choices, then how can one be responsible even for the stripped down acts of the will itself, if *they* are the product of antecedent circumstances outside the will's control? . . . The area of genuine agency . . . seems to shrink under this scrutiny to an extensionless point. Everything seems to result from the combined influence of factors, antecedent and posterior to action, that are not within the agent's control. (1979b, p. 35)

This is chilling, or so it seems. The worry has to do with thinking of every event as caused, thinking of choosing or willing as an event and thus as having causes, and, at the same time, thinking of ourselves as free agents who choose our voluntary actions in a force free field, who are "the prime movers ourselves unmoved" of what we do voluntarily.

Ever since Darwin, the idea that persons are subject to causal laws has become a regulative assumption of *Geisteswissenschaften*. And thus all the sciences—genetics, neuroscience, and psychology on one side, and sociology, anthropology, economics, and political science on the other side—can be understood as in the business of causally explaining what we are doing and why. It is not as if everyone who labors in the human sciences need be interested in causal explanation. There is a vast amount of insight to be gained from very careful thick description and taxonomizing. The point is that it is generally accepted that everything that happens in the human sphere, as in all other parts of the world, happens because antecedent things happened, and so on. Even if reasons are or can be causes (and I am sure they can be) they must have causes, and so on. Thus, it seems that the "area of genuine agency . . . shrink[s] under this scrutiny to an extensionless point" (Nagel 1979b, p. 35).

To decide whether the scientific image spells death for free will, we need to know what conception of free will is being discussed. There are two main conceptions of free will. One is the libertarian view, which is relatively new and Western. The other is older and can be found in ancient Greek and Eastern texts (I see it in Confucius, Buddha, Aristotle). The first conception cannot be fused with the scientific image; the latter conception can. So let it return.

The view that won't work is stated by Descartes this way: "[T]he will is so free in its nature, that it can never be constrained.... And the whole action of the soul consists in this, that solely because it desires something, it causes a little gland to which it is closely united to move in a way requisite to produce the effect which relates to this desire." Roderick Chisholm says this about free agency, which he calls "agent causation": "[I]f we are responsible...then we have a prerogative which some would attribute only to God: each of us when we act, is a prime mover unmoved. In doing what we do, we cause certain things to happen, and nothing—or no one—causes us to cause those events to happen." (1964)

No one has ever explained how any animal, or any natural being, could possess a part—an extremely important part, "the will"—that is not subject to causal principles but nonetheless produces astounding effects. Fortunately, there is an ancient conception of free will that has shown remarkable resiliency and that sits nicely with the scientific image.

Aristotle and Dewey

Even if there is no such thing as an incorporeal mind that governs the body, and even if there is no such thing as libertarian free will, we are owed an account of the phenomenology that supports these ideas.[31] Aristotle and Dewey are helpful here (as is the Buddha but I'll leave him out of the discussion for now). We can make peace between the scientific and manifest images in the following way: Accept that (as best we can tell) everything that happens has a set of causes that make it as it is; then proceed to distinguish the voluntary and the involuntary, the free and the unfree, in terms of the *kinds* of causation or causes that distinguish them.

Aristotle championed the voluntary-involuntary distinction long before there was a conflict between the Cartesian image of mind and agency and the scientific image. In the *Nicomachean Ethics*, Aristotle drew the involuntary-voluntary distinction this way: "What is involuntary is what is forced or is caused by ignorance. What is voluntary seems to be what has its origins in the agent himself when he knows the particulars that the action consists in." What Aristotle had in mind was something like this: An action is involuntary if it results from some sort of compulsion against which effort and thinking are impotent, or if the agent in no way knows or grasps what he is doing.[32]

Voluntary action involves the agent's knowing what action he is performing, and acting from reasons and desires that are his own. How is this possible? It is possible if I am conscious and if my consciousness has some causal efficacy. And it does. Mother Nature put me in conscious touch with some of my most salient desires, hopes, and expectations. When I see what I want and/or need and judge it to be choiceworthy, I adjust some circuitry (thanks to how I am designed, what I have learned, etc.) to do what gets me what I want. These actions are voluntary. We call actions 'voluntary' when we think about means and ends and act in accordance with our thinking (or could have).

Dewey was well aware of the conflict brewing between the libertarian (metaphysical, not political) image of persons, inspired by Descartes, and the assumptions made by the new scientific psychology. In an early paper, "The Ego as a Cause" (1894), Dewey claimed that the main question facing the science of the mind was whether we can "carry back our analysis to scientific conditions, or must we stop at a given point because we have come upon a force of an entirely different order—an independent ego as an entity in itself?" His answer was that the myth of a completely self-initiating ego, an unmoved but self-moving will, was simply a fiction motivated by our ignorance of the causes of human behavior. He saw no need for the notion of a metaphysically unconstrained will or of an independent ego as a prime mover itself unmoved in order to have a robust conception of free agency. For there to be agency, we need the person (or the ego) as a cause, possibly even the proximate cause of what we do. But the person (or the ego) may serve as the proximate cause of action and still himself be part of the causal nexus. Indeed, a person, in virtue of being a natural creature (an animal), must be part of the causal nexus.

Within five years of writing about the unacceptability of an ego as an "unmoved mover," Dewey had worked out an alternative, naturalistic account of conscious deliberation and will. In his splendid book *John Dewey and Moral Imagination*, Steven Fesmire writes:

In his 1900–1901 lectures on ethics at the University of Chicago, Dewey distinguishes four of the ways in which people deliberate, then explores the generic pattern:

(1) "Some people deliberate by dialogue."

(2) "Others visualize certain results."

(3) "Others rather take the motor imagery and imagine themselves doing a thing."

(4) "Others imagine a thing done and then imagine someone else commenting upon it."

What unifies this diversity is that deliberation "represents the process of rehearsing activity in idea when the overt act is postponed. It is, so to speak, trying an act on before it is tried out in the objective, obvious, space and time world." Following Peirce's belief-doubt-inquiry continuum, opposed tendencies create a tension, evoking an affective phase in which emotions come to the fore. This tension spurs deliberation. To escape the tension requires "an objective survey of the situation." The deliberative phase is marked by inhibition of activity and involves "turning back, as distinct from the projective or going-forward attitude." Prior experience is analyzed in order to find a way to direct current activities. (2003, p. 74)

Notice that Dewey's and Fesmire's artful descriptions of the aims and types of deliberation yield testable hypotheses about linkages between phenomenology and behavior (grist for the neurophenomenologist's mill; see Thompson 2007), about psychological development, and about character education.

Once Dewey saw the promise of a robust theory of agency without libertarian free will, he never looked back. By the time he wrote *Human Nature and Conduct* (1922), Dewey was confirmed in his opinion that, although there were big-time, large-stake questions about freedom, there were none about "free will" in the spooky metaphysical sense: "What men have esteemed and fought for in the name of liberty is varied and complex—but certainly it has never been metaphysical freedom of the will." (1922, p. 209)

Dewey is right. Cries for "freedom" are typically pleas for life, liberty, and the pursuit of happiness—i.e. for political freedom, not metaphysical freedom.

Many people think they need a notion of free agency that involves a self-initiating ego in order to undergird the idea that they are free. Maybe something else can do the job, something where the distinction between voluntary and involuntary does not turn on a distinction between acts initiated by a completely self-initiating will and those that are fully explicable in causal terms.

Dewey says the moral problem concerns the future. I treat you as an intelligent being, capable of self-control, if I call you on inappropriate or non-virtuous actions. If you are receptive and paying attention, my response gives you reason to behave better in the future. I call this *responsability* to indicate that it incorporates the credible assumption that our characters, our hearts and minds, are plastic to some degree. Social communities are dynamic systems in which complex feedback mechanisms help us adjust our beliefs, desires, feelings, emotions, and behavior.

Aristotle is silent about the source of the capacities to think and act with reason although he does say "what is voluntary seems to be what has its origins in the agent himself." But this silence should not be taken for neutrality on the existence of natural causal origins for these capacities. For Aristotle, and for many post-Cartesian thinkers, including David Hume, John Stuart Mill, and John Dewey, the assumption is that the capacities that are deployed in the initiation of voluntary action are distinctive but perfectly natural human capacities with perfectly ordinary natural histories. That the "will" seems to be self-initiating is perhaps an understandable illusion. We are not in touch, first-personally, with most of the causal factors that contribute to who we are and to what we do. It is hardly surprising that we are prone to overrate the causes we are in touch with first-personally. When I deliberate and choose among the options before me, I am in touch with the relevant processes, the processes of deliberation and choice. I am not in touch with—indeed I am normally clueless about—what causes me to deliberate and weigh my options as I do. So I make a misstep and think deliberation is self-caused. It seems that way, after all.

Spinoza diagnosed the source of the libertarian illusion this way: "Men believe that they are free, precisely because they are conscious of their volitions and desires; yet concerning the causes that have determined them to desire and will they have not the faintest idea, because they are ignorant of them." (*Ethics*, appendix to part I)

Designing persons to be in first-person touch with high-level proximate causes is a brilliant piece of evolutionary design. Were we in touch with *all* the causes of thought and action, the mind would be much too noisy, and it is doubtful that a noisy mind would enhance fitness. So a design that puts us in touch with proximate causes and screens off distal causes is fitness enhancing. But it has one unfortunate consequence. It causes us to overrate proximate causes and to think that how things seem from the first-person point of view (e.g., non-physical) is how they *are*, or, perhaps worse, that proximate conscious thoughts and intentions have no causal antecedents themselves and are produced by ourselves, where each self is a "prime mover itself unmoved."

This is as much as I will say for now about how to think of humans as very smart social animals who are agents. Agency is real, but libertarian free will is an illusion. Can we bear up and live with what scientific image says about consciousness and causation? Yes. The scientific image, if conceived carefully, need not be reductive, eliminativist, or disenchanting. If this is right, we can proceed to examine more closely what, at this point in the first decade of the 21st century, eudaimonics can say or teach about the prospects for flourishing for conscious creatures who are capable of knowing and choosing how to live in a co-dependent, co-creative journey with others.

2 Finding Meaning in the Natural World: The Comparative Consensus

Norms and Nature

We are embodied conscious beings engaged in high-stakes psycho-poetic performances. The quality of our lives, indeed whether our lives are meaningful or not, depends in significant measure on how we participate (to speak platonically) in the spaces of art, science, technology, ethics, politics, and spirituality. These six spaces are members of the Goodman set I call the Space of Meaning$^{\text{Early 21st century}}$. We cannot opt entirely out of the Space of Meaning$^{\text{Early 21st century}}$ if we want to live meaningfully. No one—not even someone who chooses to live "off the grid" and succeeds at doing so in the sense that he does not pay utility bills—can (or wants to) escape living within the constraints of a culturally available Space of Meaning.

Our nature as social mammals requires us to find meaning in a culturally available Space of Meaning or not at all. It is possible, of course, to find one's time and/or place unsatisfactory and to attempt to relocate in the past or to attempt to create new ways of being or thinking. Such relocation is commonly fantasized and sometimes succeeds. Nonetheless, relocation—backward or forward—requires moving from the Space of Meaning into which one was raised to another.

Whereas most sets are still life, the Space of Meaning$^{\text{Early 21st century}}$ is dynamic. Each member space of art, science, technology, ethics, politics, and spirituality is a complex of theory and practice. Each is layered, is nested, contains multiple members itself, and is ever-evolving, and thus so too is the whole set. Each member space differentially accumulates its own past and sheds parts of its past (possibly to re-appropriate them later). And each space interacts with itself and the other spaces in multifarious ways. The modes of interaction and intersection within each member space and across the Space of Meaning$^{\text{Early 21st century}}$ as a whole range, at any given time, from harmonious to non-harmonious. At the social level, there are reasons of communal stability to want the relations among the spaces to be toward the harmonious side. At the individual level, harmony is favored for reasons well explained by the theory of cognitive dissonance (Festinger 1957). At the group level and at the individual level, the spaces do constitutive work. Intersection with these spaces, participation in them, makes, in some measure, the community and its members who and what they are.

Given that different times and places afford different ways and means for making sense of things and finding meaning, it makes sense to ask two questions: (1) Are there any deep structural features to what humans who search for eudaimonia are looking or searching for? (2) Are there any deep structural features that measure success at achieving eudaimonia? My answer is Yes to both questions. Affirmative answers to both questions are warranted by the empirical evidence. I do not claim that our nature as social mammals provides us with a single end or goal, success in relation to which constitutes a good human life. I claim that there is a range of characteristic qualities or features of human lives that, more reliably than alternative qualities or features, link up with a good human life.[1] This might sound odd to ears trained up to worry about is-ought and fact-value distinctions, as well as about various other alleged kinds of naturalistic fallacy, such as G. E. Moore's (1903) claim that it is not possible to define the good naturalistically, because I intend questions 1 and 2 to be understood normatively. That is, I intend them as questions about how we ought to think about flourishing, about how it makes sense to think about a genuinely good life, about how to distinguish between genuine flourishing and multifarious pretenders.

It is still a fashionable philosophical dogma that normative questions cannot be addressed empirically. Consider: Should we build a bridge across the river? Why? Why not? If so, what type? This is (these are) a normative question, and I claim it is inconceivable that we would consider resolving it in any way other than empirically. Of course it is true that at some point we will have to invoke a goal or an end state. The people need a bridge in order to do business with the folk on the other side. "Where did this need, goal, end, value come from?" the philosopher wonders, as he strokes his beard with a look of eternal bewilderment on his face. The answer is this: It came from the people. Enough of them shared the goal of doing business with the folks on the other side of the river that seeking to meet this goal was judged to be a good idea. The same is true, I claim, for what seem to be, but are not, the more mysterious kind of normative questions that philosophers fuss with and often mystify. My claim is that there are substantive general things to say about what flourishing is, as well as about what brings it about. These truths and generalizations come from philosophical reflection on human nature and social life as both reveal themselves across the world over time.

The answers to both of the connected questions could have been No (or Yes/No, or No/Yes). One reason for two Nos would be if there were no rhyme or reason to what spaces of meaning there are. Whatever the spaces of meaning that make up some Space of Meaning are, they are the only locations available in which to locate meaning. Thus, if meaning is to be located, it will be in that Space or not at all. But the nature, shape, and content of that Space of Meaning, and every other one that has ever existed, is a matter of complete historical contingency. Another reason there might not be any interesting generalizations about the nature, causes, and conditions of well-being would be this: The Space of Meaning for a people is normally a functional response to a set of questions and problems they face. But the questions and problems different people face are so time-specific and context-specific that there is no deep unity that binds the Spaces of Meaning. The first case is one in which the Spaces of Meaning are random; the second case is one in which the Spaces of Meaning are functional (they have a rhyme and reason) but they are responses to random historical circumstances. In either case, there would be no interesting philosophical lessons to be gathered about eudaimonia beyond the negative lesson just stated. But there are interesting general things to say about what we seek to accomplish when we seek to flourish, and there are credible things to say about standards of success at flourishing.

Platonic Space

Insofar as we wonder about how to live meaningfully, we are locating ourselves in the wider ecology of what I have been calling 'platonic space' (because Plato was the first to try to get a grip on the orientation about which I aim to speak). The space is platonic rather than Platonic because I conceive of "the good," "the true," and "the beautiful" as natural categories, not as strict Platonic forms (Eidos) existing as immaterial phenomena. Humans across otherwise very different social, economic, and political worlds show a motivation to locate what is "good," "true," and "beautiful" in order to live by and in their lights. Living in a genuinely meaningful way has two components. First-personally one judges one's life to have been well lived *and* it is so judged by third parties in a position to evaluate legitimately the quality of the life lived.[2]

Read Plato's theory of forms in this naturalized way: "The good," "the true," and "the beautiful" are ways of gesturing at, or describing, the three fundamental and universal ways humans orient themselves in and toward the world in order to live well and meaningfully. Picture these forms initially in Darwinian terms. Mother Nature has selected for parents who care about their young, who reliably detect where food and water sources are, who choose partners with symmetrical faces (good babies), and who appreciate sunrises (work can continue) and sunsets (rest is in store). Eventually expansive containers of the multifarious things that are "thought to be" good, true, and beautiful come to exist in the social world. They are not invented by or held fully in the minds of individuals. Culture expands and articulates the forms, and each person finds his or her way to negotiate within parts of the spaces (of meaning) they provide. Ethics and politics track what is good, art and music track what is beautiful, and science tracks what is true. But this is too simple. We want the light from these forms to intermix, to interpenetrate, to harmonize—if light can harmonize. Thus we think that politics ought to produce what is good and just in an honest and truthful manner; that science which serves the good and the beautiful as well as the true is trebly wonderful, and that music, art, and literature are most powerful when they allow us to see or experience something more deeply and truthfully than any other way of worldmaking can.

On my way of reading Plato, he thought, and I agree, that in order to flourish each person must interpenetrate the spaces of the good, the true, and the beautiful to some degree. That said, there is a vast array of ways of being and living well available within the Space of Meaning$^{\text{Early 21st century}}$. And thus there are multiple ways of making sense and meaning by way of participation in the Space of Meaning$^{\text{Early 21st century}}$. However, not all ways of participation, not all paths, lead to flourishing.[3] Some do, but many don't. The general route—*dharma*, *dao*—can be made visible by close attention to the nature of eudaimonia and to the causes and conditions that reliably make it possible, or, what is different, that bring it about.

Harmony, consistency, and interpenetration of the spaces typically matter if a life is to be a good one. How do or might the components of our individual cognitive-conative-affective schemes, whereby we participate in the Space of Meaning$^{\text{Early 21st century}}$ as individuals, "hang together" so that in our search for meaning we carry out the psycho-poetic performance that is our life in a manner that is truthful, honest, worthy, fulfilling, productive of meaning, and at the same time maximally harmonious (or, at least, minimally discordant)? Assuming that this is possible to some degree, it would be good to know how we should work out and organize our own cognitive-affective-conative schemes so that we can best make sense of things and live meaningfully.

I favor speaking of cognitive-affective-conative schemes in order to mark the fact that we humans act on the basis of thoughts, beliefs, and the like, as well as emotions, feelings, and moods of multifarious kinds, all mixed in with a motivational (conative) orientation that is created over time, indeed over more time than just we as individuals have spent on Earth. What we are motivated to be or to do has a biologically based species history, as well as whatever the histories of our people and our families add to our aspirations, goals, and ideals.

Plato assumes that wisdom will reveal what these concepts—the good, the true, the beautiful—gesture at and aim to locate, and, in addition, that wisdom will enable us to see how to (a) embody all three in a consistent, unified, interpenetrating and uplifting way that (b) will constitute thinking, feeling, being, and living well—eudaimonia. One who sees what is good, true, and beautiful and who understands how they interpenetrate is positioned to flourish, to achieve eudaimonia; otherwise not.

One way—but certainly not the only way—to be a philosopher is to articulate at a synthetic and reflective level what it is that ordinary people who achieve eudaimonia, who flourish, embody and/or attain. Both the person who achieves eudaimonia and the philosopher who provides a picture of how goodness, truth, and beauty harmonize do something creative and artful.

Harry Frankfurt (1988) suggests the useful idea that what drives human meaning-making depends on the importance we assign to what we care about. Accepting my interpretation of Plato implies that there are three things people are naturally oriented to deem important and worth caring about: what is good, what is true, and what is beautiful.

One might expect two challenges:

(1) Plato didn't see that commitment to the True would yield modern science and technology. Science and technology sometimes compete with the Good and the Beautiful. Sometimes they undermine both. The Copernican and Darwinian revolutions undermined perfectly satisfying, indeed uplifting stories about humankind and its home. Plato's story of the creation of the cosmos in the *Timaeus* is revealed as poetic bunk, and the "divine quest" to live ethically and meaningfully in a way that approximates the way that is "worthy of the gods" is revealed to be a fatuous wish. We are just smart (too smart for our own good) animals. And when we die we are just dead. A really smart animal would simply go for all the gusto it can get. End of story.

(2) The claim that we have a platonic orientation is descriptively false. Many of Plato's own dialogues give voice to the still-familiar picture of persons as psychological egoists. In the famous myth of the Lydian shepherd, Plato's brother Glaucon anticipates a psychological portrait that gains rich texture with Freud. Given his or her druthers, every human wants sex, power, and whatever maximizes his or her material wealth. The sweet shepherd who discovers a ring that makes him invisible "immediately" kills the king and takes the queen and the kingdom for his own. After re-telling the myth, Glaucon says that "any man be he 'just' or 'unjust' would do the same, if he could do so with impunity." Prudence, and prudence alone, causes us to reign in this side of ourselves.

Plato and his student Aristotle have a reply: Even if we humans have motivations of this sort, and even if we were able to maximize what we see as personally profitable with impunity, we will not flourish, achieve eudaimonia, be happy. This fact, if it is one, is culled from observation of what works and doesn't work to produce eudaimonia.

One might worry that the Platonic picture is simply too sweet and wishful. Surely Thrasymachus, Hobbes, Nietzsche, Freud, and Foucault were on to something. All honest civilized souls will admit to naughty desires and urges. And we feel the force of the myth of the Lydian shepherd. The latter great minds agree that normative consensus is to be expected, but only for purely prudential reasons. We want to satisfy our libidinal urges, period. Unfortunately, a world in which each individual tries to maximize his or her own pleasure defeats itself. And thus Eros yields to Logos. Only if we are well behaved are conditions in place such that a modicum of socially acceptable pleasure satisfaction is possible. There is civilization and we are its discontents. This was Freud's last word on our situation.

There will, of course, be elaborate myth-making such as the platonic story, designed to brainwash us into thinking that we *really* are naturally oriented toward what is good, true, and beautiful and, in addition, to make normative conformity *seem* as if it provides true meaning. There will need to be, and thus there are, elaborate schemes promising amazing rewards in the hereafter, as well as art, music, and sports, and wars every so often (very often) that function to provide some subliminal satisfaction for what we would really rather being doing. But all these stories and activities are, upon analysis, just myths designed to assuage our misery—things to do to take our minds off the utter meaninglessness of it all. When Thoreau worries about all his compatriots living "lives of quiet desperation," he is simply detecting what is, indeed what must be, our predicament. Thoreau's hopefulness that flourishing could become widespread was, once again, a sweet, dear, and idle hope.

Although I may have overstated the depressing, disenchanting, "quietly desperate" picture of the human predicament, I want to grant parts of the underlying psychological picture some credence. We humans naturally care a lot about personal satisfaction. Hedonic well-being is firmly on our radar, and selfish, communally disruptive motives are abundant. It would make sense to be completely selfish—with self-serving strategic exceptions—*if* we were designed solely to achieve individual "fitness." But we were not designed that way. We were designed to be fit as *social animals*. Thrasymachus, Hobbes, et al. dramatically underestimate the degree to which social mammals such as *Homo sapiens* display what Hume called "fellow-feeling," the root of benevolence, as well as selfishness. And for all the talk about how Darwin's theory is one of "red in tooth and claw," it simply isn't. Humans are designed to care about more than individual fitness.

Hume rightly saw that which side of ourselves we reveal or which is developed more depends in some significant measure on the relative scarcity of resources. After acknowledging that our selfish sides will reveal their power in situations of extreme or moderate scarcity (the latter being the usual situation for eighteenth-century Scots like him), Hume (1777) writes:

> Let us suppose, that nature has bestowed on the human race such profuse *abundance* of all *external* conveniences, that, without any uncertainty in the event, without any care or industry on our part, every individual finds himself fully provided with whatever his most voracious appetites can want, or luxurious imagination wish or desire. His natural beauty, we shall suppose, surpasses all acquired ornaments: The perpetual clemency of the seasons renders useless all cloaths or covering: The raw herbage affords him the most delicious fare: the clear fountain, the richest beverage. No laborious occupation required: no tillage: No navigation. Music, poetry, and contemplation form his sole business: conversation, mirth, and friendship his sole amusement.... It seems evident, that, in such a happy state, every other social virtue would flourish, and receive tenfold encrease; but the cautious, jealous virtue of justice would never once have been dreamed of. For what purpose make a partition of goods, where every one has already more than enough? Why give rise to property, where there cannot possibly be any injury? Why call this object *mine*, when, upon the seizing of it by another, I need but stretch out my hand to possess myself of what is equally valuable? Justice, in that case, being totally USELESS, would be an idle ceremonial, and could never possibly have place in the catalogue of virtues.[4]

Human nature is plastic, and whether our egoistic side or our benevolent side grows more depends on a host of conditions, important among which is material scarcity or lack thereof. One might plausibly think that Hume paints too rosy a picture about Shangri-la. It might be true that if you pick the mango I want there is no cost to me in my taking the equally ripe one next to it, but if you take the woman I want there will be conflict between us. Even if sexual partners are abundant, it matters that I get the one I want. The point is that even in a situation of profuse abundance there are evolutionary reasons to think that *Homo sapiens* will naturally display aggression, jealousy, and envy in certain domains of life. There is also this possibility: a culture has formed people to be such deeply strategic self-centered thinkers that even when scarcity abates their selfishness does not.

Darwin and the Compromise Picture

Darwin's picture of human nature and the dawn of genuine sociality and eventually morality differs from Thrasymachus's and Hobbes's in two important respects. First, Darwin is a Humean, not a Hobbesian.[5] To the extent that we are egoists, we are egoists with fellow-feeling. We care about the weal and woe of, at least, some others. Second, and this follows from the first point, morality is not "something altogether new on the face of the Earth." It is not an invention *de novo*. *Homo sapiens*, presumably like their extinct social ancestors, as well as certain closely related species, such as chimps and bonobos, possess instincts and emotions that are "proto-moral," by which I simply mean that we possess the germs, at least, of the virtues of sympathy, compassion, fidelity, and courage. There is no "sky-hook" being imputed here. We didn't create the relevant instincts and emotions. Natural selection did. We are endowed with these instincts and feelings thanks to a craning operation that began with unicellular organisms. As Dennett (1995) rightly insists, if you really accept the Darwinian picture, there are no skyhooks, only cranes—cranes all the way down. A crane—a machine that is already grounded (and not itself very complex)—lifts what is already there upward and onward, sometimes making something stable and new. Skyhooks, if there are any, hang up there, on their own as it were, drawing into being that which did not exist before the skyhook drew whatever it is into being. God is a skyhook; natural selection is a crane.

What sort of cranes did nature equip us with such that morality could be hoisted? Here is Darwin's answer:

> In order that primeval men, or the ape-like progenitors of man, should become social ... they must have acquired the same instinctive feelings.... They would have felt uneasy when separated from their comrades, for whom they would have felt some degree of love, they would have warned each other of danger, and have given mutual aid in attack or defence. All this implies some degree of sympathy, fidelity, and courage.... To the instinct of sympathy ... it is primarily due that we habitually bestow both praises and blame on others, whilst we love the former and dread the latter when applied to ourselves; and this instinct no doubt was originally acquired, like all the other social instincts, through natural selection.... With increased experience and reason, man perceives the more remote consequences of his actions, and the self-regarding virtues, such as temperance, chastity, &c., which during earlier times are ... utterly disregarded come to be highly esteemed or even held sacred.... Ultimately our moral sense or conscience becomes a highly complex sentiment—originating in the social instincts, largely guided by the approbation of our fellow-men, ruled by reason, self-interest, and in later times by deep religious feelings, and confirmed by instruction and habit. (1871, pp. 498–500)

Darwin's story, as one might expect, is gradualist—gradualist in two respects. Humans, thanks largely to the possession of a cognitive-affective-conative economy that was passed on from their ancestors, have moral (or at least "proto-moral") dispositions from the start. Furthermore, these dispositions are adjustable during one's lifetime. Social insects, not being conscious, organize social life without feeling or thought. Most mammals seem to, and certainly all primates do, organize their social lives with and through feelings: selfish feelings and fellow-feelings. But if this is right, the Humean-Darwinian story has what might be perceived as a serious downside: It ties morality too closely to the emotions.

Many of the same philosophers who prefer the prettier Humean-Darwinian picture of human nature over the Hobbesian picture will say that Hobbes, even if he was wrong about what we are like deep down inside, at least saw that morality has to do with reason. Darwin, in the passage quoted above, appears to agree: Sympathy, experience, reason, instruction, and habit are all involved in the development of our moral sense. Hume, of course, thought the same. But Hume and Darwin share, in addition to a similar understanding of human nature as involving pro-social motives from the start, the view that the emotions are essential to morality even when experience, habit, and reason enter the picture. For Hume, and perhaps for Darwin, moral reason works with and through the emotions. The emphasis on the importance of the interaction of reason and emotions is therefore not a disadvantage of the Humean-Darwinian picture. First, it has truth on its side. Second, the research consensus in current psychology and neuroscience is that individuals (psychopaths and persons with Phineas Gage-like damage to medial cortical regions) who lack emotions interacting with reason are deficient at moral feeling, thinking, and action (Damasio 1994; Flanagan 2000a, 2002, 2003b).

So there is a compromise between the Thrasymachean-Hobbesian picture and a Pollyanaish one. The compromise picture is this: Deep down inside, beneath the clothes of culture, our cognitive-affective-conative constitution is a mixed bag. We have egoistic traits *and* we have traits that subserve social virtues of sympathy, benevolence, compassion, and so on.

One can imagine this objection: We have a choice between the not-so-attractive view of Thrasymachus et al. and the prettier one of Hume and Darwin. Those who favor the latter do so because it is more flattering. It is that. But that isn't the reason to accept it. The reason to accept it is that the

Humean-Darwinian picture has science on its side, namely neo-Darwinian theory. The former lacks this support.

And one might imagine a second objection: Thus far, all explanations of social virtue since Darwin's hopeful opening gambit have been instrumental or prudential. Consider the hard problem of explaining altruism. One hundred years after Darwin, with the rejection of group selectionism, everyone realized that there appeared to be no good Darwinian reasons for incurring any real costs (to fitness) to benefit any other. But such actions *seem* to occur. Two explanations came forward and have the imprimatur of the neo-Darwinian synthesis. Altruism is either (1) person-based reciprocal altruism or (2) gene-based kin altruism. The objection continues: Everyone agrees that these are the only two neo-Darwinian accounts of altruism available, and both imply the non-existence of genuine psychological altruism. Genuine psychological altruism, if there were such a thing (which there isn't), would involve incurring fitness costs for the good of the other for one's own sake, without any possibility of later reciprocation or help for one's genes. Trivers's (1971) account of "you scratch my back, I'll scratch yours" reciprocal altruism explains instrumental altruism, not any altruism genuinely motivated by the good of the other. Nor does gene-eyed altruism where I incur costs for kin because doing so might favor our shared genes. Genuine psychological altruism might occur, the critic could concede. But such cases are simply mistakes, quirky over-extensions of the imperfectly designed mechanisms that subserve the two instrumental kinds of altruism. This view is common. But it is false. Anyone paying attention knows that there are now credible accounts of how biological evolution could allow genuine psychological altruism. (The locus classicus is Sober and Wilson 1998.) Even if there emerge weaknesses in group selectionism as a realistic account for genuine psychological altruism, there is no insurmountable difficulty associated with explaining genuine altruism—as an extension of some initial setting to want to help conspecifics—so long as the relevant enhanced dispositions can be acquired. The mind/brain is plastic. And there exist great ethical traditions that teach that flourishing proceeds from altruism of wide scope. Among Buddhists everyone is encouraged to take the bodhisattva's vows to show universal compassion and loving-kindness toward all sentient beings.[6] The assumption is that expansive genuine altruism is possible, and indeed that it is actualized by many enlightened beings. And Jesus was not recommending anything he conceived as psychologically unrealistic when he recommended a Golden Rule of widest scope.[7] The point is that there are those who seek explanations of the evolution of true psychological altruism in evolutionary biology. It may be that evolutionary biology will be best at explaining kin altruism and reciprocal altruism, with plausibility proofs for how altruistic groups could also evolve (such as that proposed by Sober and Wilson). Suppose that all or most evolutionary accounts explain why altruism stays in the neighborhood, is in-group. Then they have explained why a familiar fact obtains. But our explanatory resources do not end with evolutionary biology. Given mind/brain plasticity and the transformative work that culture can do, we can explain that wide compassion and deep altruism sometimes occur, and furthermore why getting them up and running and keeping them actualized is not simple. But the fact might remain, if it is a fact, that when humans live in ways that grow their compassionate and loving sides they are more likely to find their lives subjectively fulfilling and to be judged to be living in a good way by others, often even by selfish creeps.

Psychological Wisdom, East and West

The "mixed bag" view—that we are born with an egoistic side and a fellow-felling side—is familiar not only from classic Western texts such as those of Plato, Aristotle, and the Stoics but also from other ancient wisdom traditions—Chinese and Indian philosophy, for example.

The great Confucian Mencius provides a philosophical psychology according to which humans contain four sprouts, which, if properly cultivated, lead to flourishing. The sprouts of compassion, disdain and shame, deference, and approval and disapproval, if nurtured, bear the fruits of benevolence, right action, propriety, and wisdom. If these seeds are not nurtured, we humans can turn out to be selfish creeps who do not live well or meaningfully. Mencius likens this to suffocating the sprouts of goodness. Although Mencius doesn't do so himself, one could paint a picture in which there are sprouts for *all* the traits that humans in fact reveal—the good, the bad, and the ugly. Nurturing the four good sprouts requires weeding out the sprouts that can bear the fruits of egoism and disregard for what is true. It is the job of society to care for the sprouts at first. But an individual can also learn techniques for self-cultivation and self-maintenance as he matures. Similarly, classical Indian philosophy sets out virtue, success, pleasure, and liberation from suffering as necessary conditions or constituents of the good life. These aims are clearly stated as such by contemporary Hindus. At the same time, Hinduism, Buddhism, and Jainism all agree that our natures contain hindrances, weeds, or poisons that threaten to undermine our quest to live well: egoism, greed, hatred, delusion. Despite this, all these traditions agree that true flourishing comes from abiding conventional morality, as contained, for example, in the norms embodied in the "noble eightfold path." Maximal flourishing comes from embodying the virtues of the bodhisattva: compassion, loving-kindness, sympathetic joy, and equanimity. The virtues of the Buddha and Gandhi, for example.

There are very interesting interpretive questions here about what is considered natural across various traditions. The word 'natural', as it is used in discussion of norms or ideals, obviously is not the same 'natural' that is used to mean "statistically normal." Regarding classical Chinese philosophy: Mencius did in fact grant that human beings can follow the path of gratifying their senses, rather than their moral sprouts. But nowhere does he suggest that this involves nurturing a (not-so-good) sprout. Xunzi, a critic of Mencius, does not use the sprout metaphor but argues that desires for profit and gain, and envy and hate of others are natural. David Wong (in a personal communication) writes: "It is probably a distortion of [Xunzi's] view to call them sprouts ... as he seems to have regarded only what is fully present and not latent tendencies or dispositions as part of our nature." And Hagop Sarkissian (in another personal communication) writes that "it is only Mencius who clearly employs the sprout analogy, but there is no salient notion of 'weeding out' in the Mencius text, only of growing/developing/nurturing the *good* sprouts." This suggests to me two interpretive possibilities. One is that Mencius thinks there are some not-so-good sprouts in our nature and just fails to make this clear. The other is that he thinks that bad seeds can blow in from the outside and take root in us. This would require an explanation for how, say, a corrupt culture could germinate seeds that are not in the natures of the individuals it comprises.

In classical and contemporary Indic philosophy, Hindu, Buddhist, and Jain, there is a visible strand of thinking that says that humans are innately good. I asked the Dalai Lama about this, since he says this sort of thing repeatedly in his writings. His reply was as follows: The claim that we *are* innately good (which we would think is false if intended descriptively) is a claim that describes the potential in us to achieve what is good, what is wholesome, worthy and productive of eudaimonia. Although we have seeds that support egoism (that, after all, is what the four noble truths are on about) we are not innately egotistical since that, according to the theory, would be tantamount to saying that by growing egoism we could attain what is good, wholesome, etc. And that would be false. The upshot is this: In Chinese and Indian philosophy, as in classical Greek philosophy, there is a strong strain of distinguishing among ways of living that amounts to thinking of persons and their development in botanical terms. Just as there are various sound principles to distinguish between plants that flourish, that are ripe, and so on, as well as among good seeds and bad ones (the three poisons in Buddhism of avarice, anger, and false view), there are principled ways to judge human lives to be well lived or not.

Positive Psychology and Well-Being

I have been doing serious work in comparative philosophy for more than 10 years now, first in Indian philosophy, then, thanks to my dear friends David Wong and Hagop Sarkissian, in Chinese philosophy. More recently I have been turning my eyes to Islamic and African philosophy. My convictions about the importance of making empirical assessments about how various ways of living contribute (or not) to well-being and thus to normative ethics and social and political philosophy have led me to be on the watch for what norms, if any, are universal. From the ecological and eudaimonistic perspective I endorse, if (a) there are norms that are universal as a matter of fact and (b) these norms yield practices that are productive of human good, then the best explanation (or at least a plausible hypothesis) is this: There are universal features of human nature and universal features of the natural and social environments in which we live that lead to the discovery and utilization of these norms.

There are research programs outside of comparative psychology that ask these sorts of questions. The new field of positive psychology is one such research area, and it pleases me that recent work in the field claims a comparative consensus on the virtues that are mandatory for eudaimonia. The method is to review the wisdom literature across the world's most important wisdom traditions since (say) 1000 B.C.E and see what normative consensus, if any, emerges. Here is the list of what Peterson and Seligman (2004) call the "High Six." I list them in order of alleged ubiquity:

Justice

Humaneness

Temperance

Wisdom

Transcendence

Courage.

I will talk about positive psychology and associated research programs in detail in chapter 4. Here I make two observations. First, the list conforms pretty well to my own list of universal virtues. Second, there is one notable exception: "transcendence." Peterson and Seligman remark that "'transcendence' is 'the most implicit' of the core six; Transcendence is rarely nominated explicitly, but the notion that there is a higher meaning or purpose to life, be it religiously underpinned or not, infuses each tradition" (2004, p. 50). And when "transcendence" is first introduced, they write: "The transcendent according to Kant is that which is beyond human knowledge. We define it here in the broad sense as the connection to something higher—the belief that there is meaning or purpose larger than ourselves." (ibid., p. 38)

Neither comment introducing the virtue of "transcendence," nor any of the evidence adduced give me the slightest confidence that "transcendence" so understood is a virtue. Why not? The consensus is that a virtue is a disposition to perceive, feel, think, judge, and act in a characteristic way appropriate to the virtue, i.e., as called for by a situation that makes display of that virtue, $v_1, v_2, \ldots, v_n$, apt. The first five alleged virtues fit the definition, but "transcendence" does not. The problem is that "transcendence," as described, is a general cognitive-affective-conative orientation that may well be seen the world over and may provide insight into a fundamental feature of human psychology. Furthermore, it may be normatively endorsed across cultures because being so oriented is productive of good. Nonetheless, it is not a virtue, at least not clearly so, because it is not tied to particular types of situations that call upon particular virtues. Virtues such as justice, courage, and temperance are called upon by situations that require fairness, integrity, and self-control, respectively. And they are action-guiding. A courageous person stands up for what is hers (her convictions, her body, her country). "Transcendence" describes a more global orientation to experience than virtues do. Furthermore, it is hard to see what, if any, type of action it calls forth. This point is important for several reasons. One is this: there are those who behave as if religious beliefs are immune from criticism because of religon's status as *the* primary mode of meaning-making the world over. "Transcendence" (together with impulses to make sense of things causally) is almost certainly at the root of the religious impulse to posit (a) divine being(s). But it is also at the root of certain expansive ways of feeling and thinking that deflate the self by inflating that with which the self partakes. Conceived this way, it has no necessary link to positing supernatural beings. So I think the positive psychologists are on to something. But in this case, they are not on to a virtue that is judged as mandatory for eudaimonia across cultures. The transcendent impulse is, it seems to me, very, very common. Spiritual but irreligious people display the impulse.[8] Is it good for you? I think so. But it is not, as described by the positive psychologists, a virtue.

The Upshot and Why It Matters

The general picture, which I think looks tentatively like a comparative consensus, is this: We are thrust or oriented to achieve or maximize a set of needs, impulses, or instincts only some of which will in fact bring eudaimonia. The thirsts for things that if consumed will not bring eudaimonia don't necessarily require elimination or complete repression. Minimally, we need to work to moderate and modify aims that are unproductive or destructive to our goal of flourishing as social animals. The platonic picture that sees us as oriented to locate and embody what is good, true, and beautiful must now be read as normative, specifically as a hypothetical imperative: If you want (as is wise) to achieve eudaimonia, set your sights on what is good, what is true, and what is beautiful. Why is it wise to set our sights on these things? Aristotle's answer is this: All, including hedonists and egoists, will admit that at day's end they want nothing more than to find true happiness. It is possible that if a hedonist or an egoist is convinced that hers is a legitimate way to live she might, at life's end, express satisfaction at the way she has lived. But if she has misgivings it will be because her platonic orientation still exerts a tug. The person who is eudaimon will not engage in second-guessing on her deathbed. If an egoist, a hedonist, an amoralist, or an immoralist second-guesses herself, it scores points for the view I am advocating. If this much is true, then psychology, evolutionary psychology, cognitive and affective neuroscience, and social neuroscience (Goleman 2006) need to start by accepting that we are animals possessed of the relevant cognitive-affective-conative orientation; they must then explain how and why we are constituted this way (this is not very hard; it involves the evolution of certain capacities that make us fit as social animals and that are also suited, but not designed, to satisfy urges for meaning that comes with the equipment). By starting with this view of human nature, we are positioned to describe and explain how humans as meaning-makers with certain natural orientations (toward the good, the true, and the beautiful) will predictably find themselves, individually and collectively, in situations where smooth relations among our moral, spiritual, scientific, and aesthetic orientations seem difficult to reconcile. We live in such a time. But by situating the eudaimonistic aim as the fundamentally important one we can better see both reasons for and paths back toward harmonious reconciliation among the three broad Platonic ends that, taken together, allow for the eudaimonistic aim to be realized. Although I haven't laid out the argument, I assume that the reader can see how the extrapolation works for the three platonic ends to the Goodman set with six members that I claim constitutes *the* Space of Meaning$^{\text{Early 21st century}}$. Taking the eudaimonistic aim as basic, as part of our conative constitution, is perhaps something initially seen by the Greeks in the West. It is an underestimated strain in Plato. Many classicists falsely think that Plato doesn't have much to say about eudaimonia (contrary to popular belief among many experts, the word appears numerous times in Plato's *Dialogues*) and that he focuses mainly on the metaphysical Forms and on pure rationality. For reasons that are in fact available in Plato's corpus and in Aristotle's, the descriptive and normative picture of humans as potentially fully rational animals has been ascendant in the history of Western thought. But the richer, more complex eudaimonistic strand in Platonic and Aristotelian thought has aged well, even if it has not been held so well in view. There is a growing trend among neuroscientists and philosophers—Antonio Damasio and Martha Nussbaum come first to my mind as the pioneers—to argue forcefully that a proper picture of the nature of persons requires renewed emphasis on the emotional, affective, conative, eudaimonistic thrust of our kind of animal.

Fitness, Then Flourishing

I would like to think I have given an argument that makes plausible two things: (1) Our natures are a mixed bag from a normative point of view. (2) Eudaimonia comes, if it does, only if we grow the better seeds in our nature, and weed out or moderate the growth of the "not so hot" stuff. The number of people who have—often by necessity or inattention—cultivated the bad seeds, or been given no resources to care for their own garden, are many. The facts are that their way of living does not achieve what they want, although from the point of view of a sufficiently screwed-up psycho-social-aesthetic it can *seem* pretty or good. The other way does. The generalization is based on experience. It is based on complex psycho-social-historical observation.

One way to explain how people so frequently go down a path that will not bring eudaimonia, indeed can't achieve eudaimonia, involves attending to the "first things first" imperative. Interestingly, this imperative is categorical. Fitness first!

Following John McDowell (1998), we might distinguish between first and second human nature. First nature is strictly Darwinian and comprises the conative constitution to be fit. First nature involves the powerful impulse to survive, and it engages the world with the cognitive and affective tools designed to do that job. On the cognitive side, there are pattern-recognition capacities and inductive-reasoning skills. On the affective side, there are the basic emotions. The two dimensions, cognitive and affective, are inextricably bound together from the start. Second nature kicks in once we achieve fitness or better, as we are on our way to doing so. Second nature involves something in the vicinity of the aspiration to live meaningfully, to flourish, to achieve eudaimonia.

First nature, we might say, concerns itself with satisfaction. Once we have food, water, shelter, and sex, we are impelled to seek more. What more? Happiness. The ways of making meaning and sense contained in the Goodman set that constitute the Space of Meaning$^{\text{Early 21st century}}$ have the resources to assist us in going down the right path. But we will have to wend our way through these spaces, as individuals or collectively, with our eudaimonia-detection equipment on. The spaces of meaning are cultural "objects" or sites that are there, are as they are, for all sorts of quirky reasons. They may have evolved so that we can find the good within the spaces provided, but they were not intelligently designed for that purpose. So be choosy.

One quick answer as to why so many people don't achieve eudaimonia has to do with the fitness imperative and the requirement that it be met first. Scarcity of resources keeps a full 20 percent of humans at present from achieving fitness. Flourishing can't be achieved unless fitness is.

The linear order is perhaps overstated. All the great wisdom traditions agree that certain basic needs must be met before a person has credible chances to aim for living well. Call the basic set of needs α. Even if one has α, bad luck can thwart meaning-making. Furthermore, being fit involves staying fit. Fitness is not something one achieves on a certain day at a certain time and is then done with. This means that while working to be and stay biological fit, one can begin the project of aiming to flourish. Indeed, most people do it this way.

Some critics think this view is elitist, insofar as only the 80 percent of current citizens of the Earth who are not living in conditions of absolute poverty have prospects for eudaimonia. Two points: (1) It may be true if, for example, one lives very poorly as defined by the World Bank in a society that provides minimal social support of basic necessities, even below α, that one *might* have a shot to be eudaimon. It is worth noting, however, that all examples of hermit monks involve having lived lives before going off to the Himalayan cave that equipped the hermit, sage, or bodhisattva with a way of living and being that allows continuation of the quest to live well and meaningfully. It required something like education. (2) It is not the view that is elitist. If there is elitism, it is in social practices that support doing nothing, or not enough, for those who live below the eudaimonistic threshold.

The overall picture, in summary, is this: Our innate conative constitution orients us categorically to seek fitness. This innate conative constitution is not best described as consisting of a set of beliefs—it is not epistemically driven. Nor is it, as a given in the first instance, either rational or irrational. Like induction, it is arational. We do not rationally adopt the aim to be fit or to use induction. But we know how to use the equipment we are endowed with to achieve fitness (using induction plays a big role). To use an existentialist turn of phrase, gaining fitness is the orientation we are *thrust* into the world to achieve. Next comes the magic trick: At every place one Earth, as fitness is achieved (more likely, as it is being achieved), humans begin to strive for meaning and happiness. The quest to flourish invariably reveals that we are possessed of (in virtue of the kind of social animals we evolved to be) a platonic orientation to discover what is good, true, and beautiful. Because our natures are mixed bags, this platonic part of our nature doesn't always win out. But the inference to the best explanation for how to flourish involves growing that side of ourselves. The only possibly controversial assumption I need is not really magical at all. I start with this perfectly fair and sensible question: What aspirations, if any, are quintessentially, possibly universally, human? I start this way for a good reason: I assume that all ways of making meaning and sense, especially if they are very old and, what is different, universal, reveal something about our basic conative constitution. If we have a basic conative constitution, and I think we do, it consists of dispositional setting, motivational tendencies, to seek certain ends (e.g., pleasure), to orient ourselves and our lives in a way that we think will produce satisfaction. We do things we are motivated to do before we have much sense of why we are doing them. But not much, I think, turns on this last point.

Plato described the relevant orientation, which I am aiming to tame and naturalize, in an enchanting language where the definite article precedes three holy nouns, *the* Good, *the* True, and *the* Beautiful. Experience has revealed that he was half right. We do have a constitution that aims us toward platonic space. But we now understand, post-platonically, that "the good," "the true," and "the beautiful" are not timeless external Forms but works in progress, things we seek to make determinate within the constraints of natural orientations to do so. Furthermore, and as a consequence, the words 'the good', 'the true', and 'the beautiful' gesture at what we know to be partly indeterminate, culturally variable, and, most important, multiply realizable.

But the evidence from history and philosophical anthropology leads to this picture of human nature. We are not born either fit or flourishing, but with thirsts for both. We don't experience or see the second thirst until the first is quenched. But it is there. Our conscious and unconscious minds, deploying both cognitive and affective equipment, provide the tools for achieving the twin ends of fitness and flourishing. The ends themselves are given by our natures. And thus they reveal themselves persistently over time and across space as two universal aims.

Expanding the Ecological Horizon: From Ethics to Politics

My proposal is that we think of persons as embodied beings situated in multifarious social ecologies who are driven to do as Plato and Aristotle claim: to flourish, to achieve eudaimonia. The fact that we are creatures who aim by our natures to flourish as well as to be fit is perfectly consistent with a Darwinian view. Furthermore, living in a world in which many people are prevented from actualizing their potentials places powerful moral demands on those who have the resources to do so.

Does the ecological and eudaimonistic approach I have recommended have anything to say about how our flourishing might be fruitfully conceived as tied to the flourishing of others? I think it does. John Rawls (1971) invokes what he calls "the Aristotelian Principle" in his discussion of life plans and human flourishing. The Aristotelian Principle says that, other things equal, human beings enjoy the exercise of their realized capacities (their innate and trained abilities), and this enjoyment increases the more the capacity is realized, or the greater its complexity.[9] Rawls calls this "a basic principle of motivation" that "accounts for many of our major desires, and explains why we prefer to do some things and not others by constantly exerting an influence over the flow of our activity." Now, Rawls thinks that the Aristotelian Principle is constrained by two principles of justice: (1) equal liberty for all (constrained by non-interference rules governing basic rights) and (2) "the difference principle" (according to which advantages in social and economic status are allowed only if they are to the advantage of the worst-off). Rawls claims that these two principles will be chosen by rational beings from behind a veil of ignorance, behind which no one knows where he or she will end up in the lottery of life.

Rawls attempts to speak as a careful philosophical psychologist when he writes that humans desire certain things for their own sake—"personal affection and friendship, meaningful work and social cooperation, the pursuit of knowledge, and the fashioning and contemplation of beautiful objects." "The desire to engage in complex activities rather than only simple ones," he continues, "comes from the fact that humans relish ingenuity, invention, novelty, and the expression of individuality." (1971, p. 426)

It seems to me that the Aristotelian principle and the psychology associated with it does express a fundamental fact about human psychology. It tells us something deep about the conditions for human flourishing. Meaningful human lives, we can now say, involve being moral, having true friends, and having opportunities to express our talents, to find meaningful work, to create and live among beautiful things, and to live cooperatively in social environments where we trust each other. If we have all these things, then we live meaningfully by any reasonable standard. If we have only some of them, we live less meaningfully, and if we lack all these things, especially the first two, our life is meaningless. This general framework allows that there are many, many different ways to live a good life. There are many different social habitats that can, in principle, allow for human flourishing. Furthermore, different people have different talents and interests. Which ones we seek to develop are a complex outcome of nature and nurture, as well as what our social environment favors. For many persons, realizing their complex talents and interests is not in the cards. Some are prevented by their environment from even discovering what talents and interests they have. Others know what they would do, how they would live, if they had the chance, but they don't. What Marx called "alienation" is the widespread condition of not being able to discover what one wants or not being remotely positioned to achieve it. Alienation is self-estrangement. An alienated individual feels the tug to uncover her potential, whatever it is, but her social environment prevents her from gaining the education or material support that might lead her discover who she is and what she might be. Alternatively, she may have discovered her talents and interests, but there may be no social institutions in place to realize her talents or interests. Other times, the opportunities exist, but not for her because she must devote all her energies to surviving or caring for others, possibly in squalor.

Some human lives thus lack meaning. If we think that each human life has intrinsic worth and that each person deserves equal chances to live a good life, then it follows that we should work to make the conditions of living meaningfully universally available. Aristotle saw this, at which point he turned his attention from ethics to politics.[10] From the perspective of ethics as human ecology, certain socioeconomic conditions need to obtain for humans to flourish. These requirements simply are not in place for all people. Education and meaningful work are not universally available. Beauty is hard to find amidst squalor. The lack of basic political freedoms, massively unequal distributions of wealth, and ethnic hatred provide dramatically sub-optimal environments even for basic decency and true friendship. These conditions require worldwide political reform.

What I have just said reveals that some sort of ethical and/or political principles are required if the Aristotelian Principle (AP) is to carry ethical and political weight. The AP is, after all, a psycho-social-historical generalization. Nothing normative follows directly from it. That is, without linkage to action by some sort of defensible norm or norms, the AP can just be taken to reveal a psychological feature of persons that will be pursued by each individual or, perhaps, within a small circle of friends. For the psychological insight to do additional work, it needs first, to be linked to something like the recognition that all persons are equally worthy of respect, dignity, and worth; and second, recognition that this recognition provides me with reason to act in such a way that I work to make the social conditions for achieving the AP widespread. Assume, therefore, that a meta-ecological norm, or a wide ecological norm to the effect that we are to seek human good in a way that does not privilege any group of persons, can be justified. (Kant, Marx, Rawls, Martha Nussbaum, Amartya Sen, and Peter Singer all have ways of doing this.) The project of human ecology conceived as eudaimonics then becomes seeking the world most beneficial to all human beings, including future ones.

The Capabilities Approach

Martha Nussbaum and Amartya Sen (1992; Nussbaum 1999, 2000; Sen 1987) have done important work of this wide ecological sort. They call their approach the "capabilities approach," and it is designed specifically to make judgments about how well different cultures are providing the conditions necessary for human flourishing. According to the capabilities approach, we ask what an average human is capable of achieving if she was given the chance to develop. The key question is not whether she accepts or approves of her way of being and living, but whether she has chances to develop her human capacities. Nussbaum (2000, pp. 78–79) offers a list of "Central Human Functional Capacities." This is my paraphrase:

a life of normal length

being able to have good health, adequate nourishment, and shelter

being able to move about freely and securely

being able to use sense, imagination, and thought in a "truly human way," cultivated by a good education

being able to develop one's emotions in a normal way, being able to love

being capable of forming a conception of the good, critically evaluate it, and live and plan in accordance with it

affiliation, i.e., being able to live with and toward others in a compassionate and just manner

being able to live harmoniously with nature

being able to laugh, play, and enjoy recreational activities

political *and* material control over one's environment: free participation in political life, ability to own property and possession of *habeas corpus* rights.

The reason Nussbaum and Sen emphasize opportunity to achieve the central human capacities over acceptance or approval of one's lot in life or even subjective contentment derives from two observations.[11] The first is that the Aristotelian Principle of flourishing captures the set of aims or capacities that, if satisfied, would produce a good life in an objective manner. Second, it is a well-confirmed social psychological observation that people do not always see things clearly or truthfully. Certain social systems inculcate beliefs about the stations, duties, and worths of particular persons that lead them to accept that they are not due full access to the sort of development that the Aristotelian Principle envisions. Such systems produce malformed persons, who, among other things, may fail to see what they legitimately want or need. It is worth emphasizing that everything said so far, is, I claim, based on empirical observation. The truths cited are truths of philosophical anthropology and psychology. They are truths of human ecology conceived widely, truths of a robust and defensible theory of human flourishing—eudaimonics.

I want to call attention to the fact that across all the traditions I have mentioned eudaimonia depends on ethical, aesthetic, and epistemic goods.

Either living in the vicinity of the good and the beautiful will not work to achieve eudaimonia if we believe what is false, or it will work, but deficiently so if it is supported by beliefs that can be known at the time (and are widely known by others with whom one is in relation) to be false, wishful, delusional. Similarly for only living truthfully. This is so for a variety of reasons, one of which has to do with the fact that on one conception of the true it pertains only to facts and says nothing about what is of value or about norms. If this is all we mean by "the true," it will help not one iota in answering questions about how one ought to be or to live. I do not think this is a good description of how the domain of the true was or is conceived by any great wisdom tradition, but depicting it initially in this way can help move us a step closer to understanding our contemporary situation or predicament as regards conflict among the disciplines, among *episteme* or *scientia*—organized ways of trying to see things truthfully.

Conclusion

I have deployed an empirical ecological approach to provide an answer to the question I posed at the start of the chapter: Is there anything substantive that can be said about how best to find meaning and to live purposefully, to achieve eudaimonia, given that we are fully natural beings, members of the species *Homo sapiens sapiens*?[12] I have sketched an affirmative answer in the form of a philosophical psychology that has both descriptive and normative credibility. Everything I have said is compatible with the picture of persons that emerges from neo-Darwinian theory and from the best current mind science. According to that picture, we are fully embodied thinking-feeling animals who live and achieve meaning—if we do—in a world that is fully natural. We are agents, and we act freely. But we do not possess any non-natural faculty of free will that permits circumvention of natural law. When we die, our career as a conscious being is over. But we leave effects. Our karma, good or bad, carries on. This matters. So it is wise to live well, in a way that makes meaning and sense in a manner that alleviates suffering and equips others to pursue what our common humanity makes us seek. If you live with your eye on the prize, then when you die, although you won't go to heaven, you'll have lived in a worthy way and have something to be proud of.

3 Science for Monks: Buddhism and Science

Conflict between the spaces of science and spirituality is one of the most familiar zones of conflict among the spaces of meaning that constitute the Space of Meaning$^{\text{Early 21st century}}$. The conflict between the space of spirituality, typically in its religious forms, and the space of politics is the other contender for the zone of greatest conflict. Buddhism, at least in the hands of the fourteenth Dalai Lama, provides the possibility proof that a great spiritual tradition and science can find peace, possibly even mutually enrich each other.

The Dalai Lama and Science

In several public addresses and publications, over more than two decades,[1] and most recently in his book *The Universe in a Single Atom: The Convergence of Science and Spirituality*, the Dalai Lama has said this: "My confidence in venturing into science lies in my basic belief that as in science, so in Buddhism, understanding the nature of reality is pursued by means of critical investigation: *if scientific analysis were conclusively to demonstrate certain claims in Buddhism to be false, then we must accept the findings of science and abandon those claims.*" (2005, pp. 2–3, emphasis added) This repeated statement is normally put in just this form as a sort of open epistemic welcome mat: "Come sit by my side, my Western scientific and philosopher friends. Tell me what you know. I will teach you what I know. We can debate. But in the end it is our duty, on both sides, to change our previous views if we learn from the other that what we believe is unfounded or false." But there is what Thupten Jinpa, a close collaborator and the Dalai Lama's main English interpreter, calls "the caveat." Jinpa writes:

> The Dalai Lama . . . argues that it is critical to understand the scope and application of the scientific method. By invoking an important methodological principle, first developed fully as a crucial principle by Tsongkhapa (1357–1419), the Dalai Lama underlines the need to distinguish between what is negated through scientific method and what has not been observed through such a method. In other words, he reminds us not to conflate the two processes of *not finding* something and *finding its nonexistence*. (2003, p. 77[2])

The Dalai Lama was friendly with and influenced by Karl Popper. Popper is most famous, of course, for his criterion of falsifiability: a statement is scientific just in case there are (possible) tests that could test its mettle.[3] A scientific statement that is well tested sticks its neck out, and to the degree it does so and is not undermined or falsified, it is corroborated.[4] Popper worried about claims that appear to make assertions about the natural world but that, through various techniques, immunize themselves from falsifiable tests. For the purposes of the present discussion, we ought to keep our eyes out for these two claims:

1. Mental properties are *sui generis* immaterial properties.
2. Humans die, but their consciousness continues; consciousness is subject to karmic laws of rebirth.

Why should one keep one's eye out for such claims? First, Buddhists commonly assert them, although neither is required by Buddhist philosophy as I understand it.[5] Second, one might think, given the Dalai Lama's "welcome mat" statement, that they are just the kind of assertions that *might* fall to science but which, given the caveat, could never be made to yield.

My main purpose here is to continue the profitable dialogue between Buddhism and science, emphasizing that by my lights this dialogue is a wonderful model for how respectful and profitable dialogue between science and spirituality can proceed. I focus specifically on possible differences of opinion about (1) the neo-Darwinian theory of evolution and (2) naturalism about consciousness.[6]

Since 1999, I have been fortunate to have been engaged in many discussions with Tibetan Buddhists, including the Dalai Lama, on the connection between science and Buddhism.[7] Rather than try to give a comprehensive overview of the "Mind and Life Dialogues,"[8] my strategy is to provide a critical reading of the Dalai Lama's latest book, treating it, as seems justified, as containing his considered view on "the convergence of science and spirituality."

Besides engaging personally in dialogue with Western scientists and encouraging scientific research into Buddhist meditative practices, since 2000 the Dalai Lama has led a campaign to introduce basic science education in Tibetan Buddhist monastic colleges and academic centers, and has encouraged Tibetan scholars to engage with science as a way of revitalizing the Tibetan philosophical tradition.[9] The "Science for Monks" program is itself remarkable. The consensus is that there was no internal momentum to start such a program. Achok Rinpoche (former Director of the Library of Tibetan Works and Archives), himself not previously learned in science, was the first head of "Science for Monks," an effort that involves (for now) one month of intensive training in elementary mathematics, cosmology, evolution, and mind science. In a recent interview with the Dutch scholar Rob Hogendoorn, Achok Rinpoche said: "... if you'd go [to the monastic institutes] yourself and say 'I'm going to teach science' the abbot and senior monks would probably say 'No thank you. We're not going to listen. We don't have enough time.' But when the Dalai Lama said that now is the time and that it is important to study and learn Western science, all abbots and senior geshes[10] either said Yes or kept quiet."[11]

Not only is what the Dalai Lama doing novel and not driven internally by his own Gelug tradition; history suggests that it is dangerous.[12] The Dalai Lama pays homage to Gendun Chöpel, one of the very first if not the first Tibetan geshe[13] to study and advocate the study of Western science. Chöpel is arguably the most important Westward-looking Buddhist intellectual of the twentieth century. His life ended sadly. After being imprisoned by Tibetan authorities for his communist sympathies, he spent his last years in the grip of alcoholism. Chöpel's name is to this day a source of discomfort among most Tibetan Buddhists for his dalliance with science, communism, and modernity and for his critiques of orthodox Buddhist epistemology (Lopez 2005).[14] On the other hand, the Dalai Lama is a wise man, and it has not been lost on him that there is widespread interest among scientists and philosophers in Buddhist philosophy. Einstein, most certainly an atheist by Abrahamic standards, is not the only scientist to notice that, if there is to be conciliation between science and spirituality, no spiritual tradition is better positioned than Buddhism to participate in the conciliation. The Abrahamic religions, Judaism, Christianity, and Islam, are often described by Western theologians as the apotheosis of religion, finally shredding paganism, pantheism, polytheism for belief in the One

True God who is the same everywhere: Yahweh, God, Allah. But modern science and theism—especially the sort that claims to be in possession of texts written by God—don't, as it turns out, sit together comfortably. Buddhism, being intellectually deep, morally and spiritually serious, but non-theistic and non-doctrinal, sits well poised to be an attractor for the spiritually inclined.[15]

Of course, one might imagine someone saying "Why care about conciliation? Science now holds all the cards." This is false. I argued in the previous two chapters that the way around catching the disease of finding science disenchanting is to acknowledge (a) that science hardly answers all legitimate questions and (b) that humans have natural and worthy impulses to locate the good, the true, and the beautiful and to find meaning by living in their vicinity.

In any case, the Dalai Lama believes that science and Buddhism share a common objective: to serve humanity and create a better understanding of the world. This is probably too generous with regard to the impulse behind science. Buddhist philosophy starts *all* inquiry with the intention that it benefit *all* humanity, possibly *all* sentient beings. Many scientists will claim to be in search of *the* truth, regardless of consequences.[16] Truth be told, the aims of scientists and technologists are multifarious. The knowledge-for-its-own-sake school of thought is common among, say astrophysicists and elementary-particle physicists, who see no clear practical consequences for what they discover. On the other hand, these scientists, and those who follow their work, often comment on the ways such work uncovers great beauty. It serves both "the true" and "the beautiful." On the other side, there any many practically minded scientists who seek knowledge that will be, at least in the short term, useful, but often only for individuals or countries with the financial resources to pay for what they discover. For every scientist and engineer who works in water science, agronomy, or infectious diseases, there are as many who focus on making life "better" with technological gadgetry for those already living very advanced lives.

That said, science offers powerful tools for deepening human understanding of the interconnectedness of all life, although not all scientists see or avow this goal. Despite the actual heterogeneous motivations among scientists, it is not an idle hope that in good hands scientific knowledge can enhance wholesome, ethical goals and can lead to action that benefits all sentient beings and the environment.[17] The Dalai Lama summarized these ideas in his 1989 Nobel Peace Prize acceptance speech:

> With the ever-growing impact of science on our lives, religion and spirituality have a greater role to play reminding us of our humanity. There is no contradiction between the two. Each gives us valuable insights into the other. Both science and the teachings of the Buddha tell us of the fundamental unity of all things. This understanding is crucial if we are to take positive and decisive action on the pressing global concern with the environment.

Life and Mind

In the remainder of this chapter, I attempt to advance the dialogue between Buddhism and science by discussing two topics of considerable importance: (1) Buddhism and the theory of evolution and (2) Buddhism and mind science. I choose these two foci because the nature and origins of life and mind are hot-button issues in the West and because, despite the extraordinarily polite dialogue between Buddhism and science, these are two areas of controversy. Indeed, possibly because of the politeness of the dialogue thus far, there have been lost chances for debating Buddhists on certain matters. Buddhists are trained to debate, so this is unfortunate.[18]

In a review[19] of the Dalai Lama's book *The Universe in a Single Atom*, George Johnson suggests that there are shadows of intelligent design lurking in the text, and that, in addition, there is no doubt that the Dalai Lama, despite his close collaboration with neuroscientists, thinks that the immateriality of mind is hardly ruled out scientifically.

My own view is that the proponents of intelligent design, at least in the form that aligns most nicely with Christian beliefs about an omniscient, omnipotent, and benevolent God, will find no support in the Dalai Lama's writings on evolution. One reason is that the Dalai Lama is barred by his own anti-theological views (which include finding no credibility in cosmological or design arguments) from thinking that complexity in evolution is best explained by the occasional interventions of an intelligent designer or a team of them. Regarding mind, it is true that immaterial mental properties are not completely ruled out by mind science. But the inference to the best explanation is that there are no such things. The reason has to do with mental causation. (See chapter 1.) If mental events (for example, intentions to act) are, as they seem, causally efficacious, then the best explanation is that they are neural events (neurophysicalism). Mental transformation of mind by mind is best explained as a form of downward causation by a complex, subjectively controlled psychological economy that allows the mind/brain-in-the-body-in-the-world-with-a-history to adjust, modify, and change itself. One can hold on to the view that some or all mental events are disembodied (that is, immaterial), but only, as I see things, at too high a cost. One will have to embrace some form of epiphenomenalism (the view that mental states lack causal efficacy). In 1890, William James called epiphenomenalism an "unwarrantable impertinence" in view of the state of psychology. From where I stand, it still seems so.

Philosophical Foundations of Buddhism: No God, No Self

Buddhism originated in 500 B.C.E. when Shakyamuni Buddha, also known as Siddhartha Gautama or simply Buddha, gave his inaugural address at Deer Park, near the outskirts of Benares, India (now called Varanasi). Depending on how one understands the orthodox Vedic or Indic spiritual tradition of that time, Buddhism is either a complete break with that tradition or a development of it.[20] Buddhism rejects the caste system on ethical grounds. More interesting to those who think of religion as requiring belief in divinity, Buddhism rejects both the idea of a creator God *and* an immutable, indestructible soul on logical and empirical grounds.

Insofar as the reigning orthodoxy conceived of Brahman as the prime mover itself unmoved, Buddhism rejected that idea.[21] It also rejects the idea that each individual houses an unchanging self or soul (*atman*). Beyond this, many familiar Indian ideas, including the deep importance of the appearance-reality distinction, the idea of reward for virtuous action (*karma*), the idea that suffering (*dukkha*) defines the human predicament (*samsara*) and that liberation (*nirvana*) or enlightenment (*bodhi*) is possible through wisdom (*prajna*), concentration (*samadhi*), and virtue (*sila*), and the ideas of reincarnation or rebirth, are all retained and developed in Buddhism—although, in certain quarters, with hesitancy.

Let me stick with the two metaphysical beliefs that Buddhism rejects: a creator god and a permanent self or soul. First, Buddhism sees right through the familiar problems with cosmological and design arguments for the existence of God. Such arguments beg the question of the origin of the creator or designer. To say that the prime mover always was, or is self-creating and self-sustaining, is to accept the infinite regress of causes (this one a *causa sui*) that such arguments are designed to make evaporate, which they reject as a possibility.[22] If God always is and shall be, then God itself is infinitely regressive.

When the Dalai Lama listens to the story of the big bang occurring 14 billion years ago, he says "But not, of course, the first big bang." This response is hardly a rejection of our theory of the big bang. The Dalai Lama sees the big-bang theory as itself inadequate because it is not deeply causal enough. Some scientists themselves are now wondering if a better story doesn't involve less of a singular, original bang than an origin for *this* universe that involves an open worm hole from another "parallel universe," with these other universes or their ancestors being beginningless.[23]

Cosmologists sometimes say that one can't ask what there was before the singularity banged or how it got there. What they mean is that time as physics understands it begins with the bang. But this hardly makes the sense behind the question go away. Thus other cosmologists will admit the legitimacy of the question and say they have no clue as to how to answer it. Buddhism is comfortable with an infinite regress of natural causes. Indeed, the idea fits well with the metaphysical idea of dependent origination according to which everything that happens depends on other things happening.[24]

The rejection of the Vedic (Indic) doctrine of *atman*, the idea that humans are possesses of an immutable, indestructible self or soul, comes from two lines of thought.[25] First, there is the idea of dependent origination that I just mentioned. Everything is in flux and all change is explained by earlier change. The principle is universal and thus applies to mind. Next bring in experience or phenomenology: one will see that what one calls 'the self' is like many other natural things partaking of certain relations of continuity and connectedness. My conscious being is much more streamlike than it is like Mount Everest (which is also part of the flux, just less visibly so). Conventional speech allows us to re-identify each person by her name as if she is exactly the same over time. But in fact identity is not an all-or-none thing. Personhood is a work in progress. This is the doctrine of *anatman* (not-self). Properly understood, the view is not nihilistic. One of my students once asked, in a very disturbed manner, "If I am not myself, who the fuck am I?" I am happy to report that further therapy about the meaning of the doctrine of *anatman* calmed him. Indeed, in the West a very similar view is widely held from Locke on (Flanagan 2002). And it fits nicely with contemporary mind science. Furthermore, the doctrine of *anatman* suits Buddhist (and perhaps, in an underestimated way, Hindu) ideas that persons can in fact transform themselves, become enlightened, and so on. If one's essence is, as it were, immutably fixed, it is hard to see how self-transformation is possible.

Buddhist Epistemology and Scientific Epistemology: Experience, Reason, and Texts

I have already noted that the Dalai Lama does not enter the dialogue in a completely neutral way. All inquiry ought to be undertaken so as to contribute to the alleviation of suffering. This is a proper, non-negotiable, expression of what Buddhist ethical inquiry has led to as the highest aspiration.[26] With this aim and this aim alone we express our commitment to improving the universal existential predicament of sentient beings: we seek to flourish, individually and collectively, but there are features of the world and our natures or the two in interaction that present obstacles and that need to be moderated, modified, and overcome as far as is possible.

One reason for hopefulness about the prospects for mutual integration of Buddhism and science comes from standard Buddhist methodology which the Dalai Lama describes this way: "Buddhism accords greatest authority to experience, with reason second, and scripture last." (2005, p. 22) This is indeed an auspicious starting point for the dialogue. As the Dalai Lama goes on to describe what may be the biggest difference between science and Buddhism, he focuses on the status of the scriptures: "By contrast with religion, one significant characteristic of science is the absence of an appeal to scriptural authority as a source of validating truth claims." (p. 25) Here I think he may underestimate the status and the frequency of appeals to authoritative "scriptures" in science. First, unlike the Vedic tradition of India or the Abrahamic traditions in the West, Buddhist scriptures are scriptures that are decidedly *not* thought to be divinely authored. They are themselves accumulations of cultural wisdom organized and authorized as authoritative because the wisdom they contain passes tests of experience and reason. It can be argued that science has scriptural traditions of its own that function in pretty much the same way. When Newton spoke of "standing on the shoulders of giants," he was acknowledging his dependence on prior science. Had Newton lived when Ptolemy or Copernicus did, he might, being very smart, have come up with their theories, but he could not have come up with Newtonian physics without the antecedent work of Copernicus, Galileo, Kepler, and others.

Both Descartes and Spinoza make much of the fact that in order to know we all must trust our ancestors deeply.[27] Spinoza points out that our knowledge of parentage and the date of our births is, in fact, "knowledge by hearsay." My students always find this a novel thought, amusing and true. Descartes' own epistemological crisis that results in his resolution to "methodically doubt" all his previous knowledge is motivated by the recognition that nearly everything he believes[28] comes from his sensory experience or from his having heard about and learned what the great minds of the past discovered.[29]

But there is one difference that makes a difference regarding the legitimacy of past wisdom. Science requires that the accumulated wisdom of its own past be intersubjectively re-testable, where this means that anyone with the suitable instruments can replicate the results should they come into question. Buddhism distinguishes among enlightened and unenlightened beings. This allows certain knowledge that (say) Nagarjuna writes about to remain in place, authoritative, even if I can't confirm it by my own experience and reason. Why? Because he is judged by wise and critical readers to be very enlightened—more so than the rest of us.

In any case, the Dalai Lama enters the dialogue with great hope because of a certain epistemic common ground. But he also enters the dialogue with a legitimate concern about what he calls "scientific materialism": "I have noticed that many people hold an assumption that the scientific view of the world should be the basis for all knowledge and is knowable. This is scientific materialism.... It is difficult to see how questions such as the meaning of life, or good and evil can be accommodated within such a worldview. The problem is not with the empirical data of science but with the contention that these data alone constitute the legitimate ground for developing a comprehensive worldview or an adequate means for responding to the world's problems." (2005, pp. 12–13)

We can do a certain amount of therapy for all participants in the debate by saying that Scientific materialism comes in a tame, methodological form and in an extravagant, imperialistic form. The tame version (call it scientific materialism^{tame}) says that, when it comes to trying to understand what there is in the natural world and how it works, you should use only the best scientific epistemology and ontology to regulate your inquiry. The imperialistic thesis (call it scientific materialism^{imperialistic}) receives ontological expression this way: What there is, and all there is, is what science says there is. Its receives epistemological expression in a form such as "insofar as there are truths humans can know, they can only be known scientifically," or (relatedly but not equivalently) "everything worth expressing can be expressed scientifically." I prefer to call these later views global metaphysical materialism and scientism, respectively. Together they constitute scientific materialism^{imperialistic}, a silly and puffed-up view that is not worth taking seriously. It is attacked by many wise people, including McDowell (1996) and Wallace (2000).[30] I am not sure I know anyone who holds the view. But I will make believe it is not simply a bogeyman and show how to defeat it, bury it, and remove it permanently from the discussion.

It is remarkably easy to make quick work of the imperialistic views. First, scientism[31]: Science has no language to express "oughts," unless perhaps they are the "oughts" of engineering or those of technological practice. If it is true that one ought to tell the truth, then there is a truth that is not expressible scientifically. Furthermore, if this and many other moral truths are knowable, then there is knowledge that is not scientific. Regarding ontology: The claim that what there is, and all there is, is what science says there is either is or isn't a claim of science. But it can't be a claim of science, since (putting aside what it even means) no such universally quantified claim has been confirmed scientifically. The kindest interpretation of the global claim is that it is a non-scientific regulative ideal for what, at the end of inquiry, metaphysics will countenance as the ultimate fabric of the universe. But we are not at the end of inquiry, so the claim starts to look like a silly advertisement for we know not what. In either case, the ontological claim, hope, regulative ideal, or advertisement is not itself a piece of science, so it cannot be sensibly asserted by the proponent of the view.

In any case, I have already noted that the Dalai Lama says that Buddhism will yield to science if there is incontrovertible evidence against some claim of Buddhist philosophy. I don't think he means to include Buddhist ethics in the class of commitments that might be adjusted. It is too well confirmed by experience and reason, and it has thus achieved canonical status in the scriptures. He is thinking of Buddhist cosmology, psychology, philosophy of mind, and metaphysics.[32]

Evolution

The Dalai Lama and his colleagues have been exposed to and studied the basics of the neo-Darwinian theory of evolution.[33] What reservations, objections, or misgivings, if any, do they have about that theory? My strategy is to claim that what can seem to be a disparate set of concerns mostly revolve around differences in views about causation, specifically, around whether there is one or two kinds of causation.

Darwinian theory is about the processes that govern the evolution of life. It is not a theory of everything, or even a theory about all aspects of living things.[34] For the first 10 billion years after the big bang, the universe expanded. Let us imagine that it was only on Earth at least 6 billion years after the big bang in an environment of carbon, methane, nitrogen, ammonia, water, and electricity that protein structures began to form eventually giving rise to unicellular life. At this point the laws of evolution by natural selection began to do their thing. Between 4 billion years ago and now, many species formed—including, eventually, *Homo sapiens*, perhaps 100,000 years ago.[35]

"My own view," the Dalai Lama writes, "is that the entire process of the unfolding of a universe system is a matter of the natural law of causality." (2005, p. 90) Put this way, his position is perfectly consistent with the combined story of physics, chemistry, and biology. But then he makes it clear there is another kind of causality: karmic causality. Whatever karmic causality turns out to be, the idea is not crazy. After all, according to the picture of the unfolding universe I just presented there is emergence. Inorganic stuff governed only by the laws of inorganic chemistry gives rise to organic material governed by the laws of organic chemistry; eventually there is life, at which point the laws of biology and evolution come into play.

We need to understand what 'karma' means, and then what 'karmic causation' means and how (if there is such a thing) it works. 'Karma', the Dalai Lama writes (2005, p. 109), "means 'action' and refers to the intentional acts of sentient beings. Such acts may be physical, verbal, or mental—even just thoughts or feelings—all of which have impacts upon the psyche of an individual, no matter how minute. Intentions result in acts, which result in effects that condition the mind toward certain traits and propensities, all of which may give rise to further intentions and actions." All this makes perfect sense. It is nothing a philosophical naturalist will puzzle over. The actions of sentient beings have all sorts of effects, some of which, insofar as they involve reproduction, reproductive rates, and positive and negative environmental effects, have consequences for fitness. Social and political effects may have no consequences for fitness but may affect the quality of our lives and those of future generations.

The Dalai Lama continues: "I envision karma coming into the picture at two points. When the universe has evolved to a stage where it can support the life of sentient beings, its fate becomes entangled with the karma of the beings who will inhabit it. When we use the term 'karma', we may refer both to specific and individual acts and to the whole principle of such causation. In Buddhism, this karmic causality is seen as a fundamental natural process and not as any kind of divine mechanism or working out of a preordained design. Apart from the karma of individual sentient beings, whether it is collective or personal, it is entirely erroneous to think of karma as some transcendental unitary entity that acts like a god in a theistic system of a determinist law by which a person's life is fated. From the scientific view, the theory of karma may be a metaphysical assumption—but it is no more so than the assumption that all of life is material and originated out of pure chance."

Again, so far so good: The idea can be understood straightforwardly as follows:

(1) Once sentient beings exist, they think, feel, and act in ways that have effects. These effects are of two kinds: personal, both intrapersonal (on the person himself) and interpersonal (on those with whom the person interacts) *and* environmental, affecting the natural and built world. To these one should add social, economic, and political effects.

(2) Karmic causation as depicted in (1) is natural. It is not due to theistic intervention at the beginning of the process (say, in creating a big bang with a plan), nor is there intentional (intelligent) design along the way other than the effects of the sentient beings (human and nonhuman) who are creating karmic effects = effects via their actions.

Interpreted in this tame way, it is not clear why the Dalai Lama says that "from the scientific view, the theory of karma may be a metaphysical assumption—but it is no more so than the assumption that all of life is material and originated out of pure chance." There is no reason for defensiveness. As I have interpreted the Dalai Lama's explanation, karmic causation is a well-confirmed fact. Once there are sentient beings who can perform intentional actions, they create all sorts of effects. The very idea of human sciences (*Geisteswissenschaften*) is based on this premise. In both cases, what is being called an assumption is of an altogether different order than my assumption that Duke University will win the national championship in basketball this year. That assumption is akin to a guess based on hope, loyalty, and a certain amount of inductive evidence. But the odds, alas, are that my assumption will not pan out. There are too many worthy competitors, and Duke's current team is the youngest since the Second World War.

The situation is different with the two assumptions about causation. Each assumption—(1) that there is ordinary causation and (2) that sentient-being causation is an interesting subtype—began as an inductive generalization that was then confirmed again and again, eventually becoming regulative assumptions of *Naturwissenschaften* and *Geisteswissenschaften*, respectively. Both assumptions keep working, keep being confirmed. So, speaking in a Kantian idiom, both assumptions now have something like constitutive status. They began as plausible inductive generalizations or inferences to the best explanation on the basis of the initial data, and they have been corroborated again and again.

But now some more puzzling ideas surface. The Dalai Lama writes: "More difficult perhaps is the first intervention of karma, which is effectively the maturation of the karmic potential of the sentient beings who will occupy that universe, which sets in motion its coming into being." (2005, p. 110) Let us parse this sentence into two. First there is the question of when karmic causation began to operate. From a naturalistic perspective, and specifically from an evolutionary and mind-science point of view, the answer is this: Karmic causation began whenever sentient beings began to act and leave the effects they leave. But the Dalai Lama puts what could be the very same point, but probably is not, in this puzzling way: The first intervention of karma "is effectively the maturation of the karmic potential of the sentient beings who will occupy the universe, which sets in motion *its* coming into being." This is hard to interpret. To what does 'its' refer? It could refer to the time when (actually the long period during which) sentient beings evolved, became abundant, and started affecting the trajectory of the world in dramatic ways. No one knows exactly when this was. But that it did happen is accepted by Darwinian theory. How it happened is much more puzzling. Among philosophers and scientists who accept the neo-Darwinian theory and thus accept that sentience evolved from insensate life (bacteria for whom "there was nothing it is/was like to be them"), there are only stories, some more credible than others, as to why evolution favored an engineering solution that, across many distinct lineages, produced sentience. No really good view of how and why (these being different questions) it happened has been satisfactorily worked out. Nonetheless, for the naturalist any story worth entertaining must conceive things in a way that is possible according to the lights of our best evolutionary theory (or theories) (Flanagan 1991; Flanagan and Polger 1995; Polger and Flanagan 2002). This, of course, is a pretty weak constraint. That said, the Dalai Lama's statement could be interpreted as a controversial answer to the question of why sentience emerged. He could mean something like this: At one point the universe harbored incipient or unactualized sentience. Still, one might say the Darwinian is committed to much the same idea: Whatever did emerge, "sentience," in the case at hand, is the actualization of the potential of antecedent natural processes. To create genuine controversy here, one must add this: When sentience emerged, it was *not* due to Darwinian processes. It involved instead "the *maturation* of the karmic potential of the sentient beings who *will* occupy that universe." The maturation of this potential sentience "sets in motion that universe's [the one with sentience in it] coming into being." Karmic causation^{untame} now sounds suspiciously teleological and, at the same time, seems to involve something beyond karmic causation^{tame}.

Having come upon what seems like a difference of opinion about matters of importance, we are now at a point where we can understand something about how differences between the epistemology of science and Buddhist epistemology might be in play at this juncture. I just acknowledged that there exists no widely accepted story about why sentience evolved—why, for example, we didn't evolve as just very smart information processors. Furthermore, at this time no one has a clue how to perform experiments that would show how behaviorally equivalent zombies and conscious creatures would fare against each other in a fitness challenge. That said, Darwinian theory is extraordinarily well confirmed. Work with fruit flies can let us watch Darwinian principles at work over very short periods of time, as can everyday work in agriculture science. So those who say that Darwin's theory is only retrodictive, untestable, and the like are very seriously misinformed. But as regards the evolution of sentience, we are pretty much, at this time, left with "just so" stories. Darwinians revert to their received wisdom: It happened, so it must have happened in a Darwinian way. The Dalai Lama reverts to past Buddhist wisdom. In both traditions, when experience and reason lose their grip, one goes to received wisdom. Thus the Dalai Lama asks:

> How do Buddhist cosmological theories envision the unfolding of the relationship between the karmic propensities of sentient beings and the evolution of a physical universe? What is the mechanism by which karma connects to the evolution of a physical system? On the whole, the Buddhist *Abhidharma* texts do not have much to say on these questions, apart from the general point that the environment where a sentient being exists is an "environmental effect" of the being's collective karma shared with myriad other beings. (2005, p. 92)

This is not an unfair statement of the situation of the evolutionist when it comes to explaining the evolution of sentience. The evolutionist will, as I have said, nonetheless insist that sentience evolved in a Darwinian manner. Why? Because the theory of evolution is very, very well confirmed. It is the best—really the only—well-confirmed theory that makes the emergence of sentience in the biological world not completely mysterious. In the first instance, creatures who detected noxious and good-making/survival-enhancing stimuli by pleasant and painful experiences appeared. This turned out to enhance fitness. And so creatures with the capacities to have "subjective experiences" proliferated. Eventually we, *Homo sapiens*, came to be. Thanks to sentience and smarts we did really well. Now there are more than 6 billion of us. The increasing weight (literally, one might say) of sentient beings-in-the-world increases the amount and complexity of the karmic effects of these beings. And this affects the way the world is, the way it is unfolding, and the prospects for all things and beings.

The point is that the Dalai Lama's wonder or worries about how ordinary and karmic causation fit together, if understood in the tame way I have recommended, are not cause for any worry from a perspective that binds neo-Darwinian theory to mind science or to the human sciences—*Geisteswissenschaften* conceived without *Geist*—more generally. Sentient creatures are part of the biological fabric of the universe, thus whatever causal effects they produce affect the world. All the sciences taken together are in the business of explaining how the relevant causal features and factors contribute to the unfolding of us and our world. It seems though that this might not satisfy the Dalai Lama, since he adds this: "The ability to discern exactly where karma intersects with the natural law of causation is traditionally said to lie only within the Buddha's omniscient mind." (2005, p. 92) Here, if the Dalai Lama is endorsing seriously the idea of consulting "the Buddha's omniscient mind," it is a difference that makes a difference between the epistemic rules in science and in Buddhism for how, using their respective "scriptures," we are to resolve puzzles that experience and reason don't resolve. In science, even if experience and reason can't yield a definitive test for some unsolved problem, the rules require trying to gain some explanatory grip by looking to the best intersubjectively confirmed theory in the vicinity, the theory with the best potential resources for making sense of the puzzling phenomena. However, in Buddhism, because some minds are thought to be more enlightened than others, they may see or know solutions that are not intersubjectively available.

One might politely point out to the Buddhist that saying that the solution to the problem confronting us is known only by the "omniscient Buddha" is really just a way of saying that the problem will remain eternally mysterious for us non-omniscient souls. But there are two reasons for greater optimism. First, no real Buddha is omniscient; he is just very enlightened.[36] Second, the problem of reconciling a world in which no sentience exists and in which ordinary physical, chemical, and biological causation do their things, and the world that emerges when sentience and karmic causation co-emerge, is not that difficult unless one introduces either of two theses that will cause serious explanatory obstacles: (1) that the emergence of sentience was planned or metaphysically pre-ordained; (2) that sentient beings are not animals, specifically, that consciousness is of an ontological type that does not abide natural laws.

Invoking (1) and/or (2) is not required by the distinction between ordinary causal principles and karmic causation (where the latter are a subtype of ordinary causal principles), nor by anything in my analysis of the Dalai Lama's views so far. Remember that the Dalai Lama writes: "By invoking karma here, I am not suggesting that according to Buddhism everything is a function of karma. We must distinguish between the operation of the natural law of causality, by which once a certain set of conditions are put in motion they will have a certain set of effects, and the law of karma, by which an intentional act will reap certain fruits." (2005, p. 90) This seems innocuous enough. But enough is said in the Dalai Lama's book, and enough is known independently about Buddhist views on rebirth and the nature of mind, to think that the contrast between ordinary causation and "the law of karma, by which an intentional act will reap certain fruits" means to say, or imply, something controversial.

What I have been calling the tame interpretation of karmic causation involves the conjunction of three uncontroversial ideas: (i) that sentient beings exist; (ii) that these beings engage in mentally initiated purposeful action, and (iii) that all the actions of sentient beings (intentional and unintentional) have abundant effects. If (i)–(iii) are uncontroversial, one might wonder why it is worth distinguishing karmic causation from ordinary physical causation at all. One reason is this: Even if sentient-being causation (= tame karmic causation) is not ontologically distinctive, it is epistemically and explanatorily interesting and informative to mark it off as a distinctive subtype. It depicts the causal intricacies of the lives of sentient beings, especially when they act intentionally, in the right way.

According to the theory of evolution, sentience is a biologically emergent feature of non-sentient biological life. Sentient beings constitute a subset of living things. Vegetables and unicellular organisms are alive but not sentient. Sentient beings that can consciously control their own thoughts, feelings, and actions are a subset of sentient life. Some think that the relevant powers are possessed only by *Homo sapiens*; others think that all mammals, possibly some or all birds, amphibians, even reptiles possess the relevant capacities to some degree, at least the capacity to act to get what they want. Thus the idea of karmic causation makes sense, at the right level of analysis, of the operation of sentient beings. Understood in this way, it could just be called 'sentient-being causation'—a sub-type of ordinary causation. Karmic causation understood as sentient-being causation gives *Geisteswissenschaften* the conceptual distinction it needs to attend to the operation of the part of the world it is designed to explain at the right level of analysis.

But the less tame interpretation would run like this: Karmic causation^{untame} is intended to do more work than expressing (i)–(iii). It names an ontologically unique kind of causation that accounts for how the psyches of future beings are determined by a set of causal processes that involve more than the environmental plus psycho-social-political-economic effects of previous occupants of the Earth. What is meant by the idea of "the law of karma, by which an intentional act will reap certain fruits" is this: my consciousness does not die when my body does, it goes on and reaps in the next and possibly many (many) future lives what it sows in each antecedent life. This, one might say, is simply what all eschatologies say in one form or another. So in order to make clear what makes karmic causation^{untamed} especially distinctive, one might add: immaterial properties of sentient beings produce causal effects in the natural world, upstream, down the road, in the future.

On both the tame and the untame interpretation, the world evolves as it does, in some significant measure, due to the effects of how humans live. But according to the law of karma (untame) the actual psyches of future beings are juiced by a karmic reward and punishment system. This is the crux, I think, of why the Dalai Lama expresses misgivings with both the randomness and lack of directionality in evolution: "As the American biologist Ursula Goodenough aptly put it at a Mind and Life conference in 2002, 'Mutation is utterly random, but selection is extremely choosy!' From a philosophical point of view, the idea that these mutations, which have such far-reaching implications, take place naturally is unproblematic, but that they are purely random strikes me as unsatisfying. It leaves open the question of whether this randomness is best understood as an objective feature of reality or better understood as indicating some kind of hidden causality." (p. 104)[37]

I interpret the "hidden kind of causality" referred to here as an aspect of karmic causation^{untame} that gives direction to future worlds by way of the system of laws that govern karmic payoffs in future rebirths. Something more than the efficient causation warranted by the concepts of ordinary causation and karmic causation^{tame} is being introduced. And non efficient causes are looked upon with suspicion by science.[38] If this is correct, or in the right vicinity, then the caveat mentioned at the beginning comes into play. Jinpa, the Dalai Lama's close collaborator and English interpreter, states the caveat this way:

The Dalai Lama, however, offers an important caveat. He argues that it is critical to understand the scope and application of the scientific method. By invoking an important methodological principle, first developed fully as a crucial principle by Tsongkhapa (1357–1419), the Dalai Lama underlines the need to distinguish between what is negated through scientific method and what has not been observed through such a method. In other words, he reminds us not to conflate the two processes of *not finding* something and *finding its nonexistence*. (2003, p. 77)

The applicability of the caveat, in the present context, is straightforward: Science finds no evidence for rebirth, but it has not found its "nonexistence." It follows according to the spirit of the caveat that belief in rebirth is acceptable. It is not as if it is simply acceptable because it is part of the Buddhist tradition to believe it. The Dalai Lama says that the great epistemologist Dharmarkirti "clearly did not think that the theory of rebirth was purely a matter of faith. He felt that it falls within the purview of what he characterized as "slightly hidden" phenomena, which can be verified by inference" (2005, p. 133). A comprehensive and responsible critique of the doctrine would need to look at the quality of the inference, which I won't undertake here.[39] (See Willson 1987.)

However, I will say more about the meaning and the epistemic status of the caveat as I now bring the final (for my purposes) topic for discussion more clearly into view. This final topic involves the nature of the conscious mind, mental properties, and mental events.

The possibility that consciousness is ontologically independent of natural processes lurks in our discussion of karma, karmic causation, and the prospect that consciousness does not die when I do, or when "my body" dies. It is time to merge the discussion of evolution with the dialogue as it pertains to the nature of mind.

Matter and Consciousness

The Dalai Lama clearly ties the two problems together in this passage:

The problem is how to reconcile two strands of explanation—first, that any universe system and the beings within it arise from karma, and second, that there is a natural process of cause and effect, which simply unfolds. The early Buddhist texts suggest that matter on the one hand and consciousness on the other relate according to their own process of cause and effect, which gives rise to new sets of functions and properties in both cases. On the basis of understanding their nature, causal relations, and functions, one can then derive inferences—for both matter and consciousness—that give rise to knowledge. (2005, p. 91)

It will be useful to distinguish three major philosophical issues that cause concern for those of us who want to understand the conscious mind: (1) the question of the role of phenomenology in the study of the mind, (2) issues of epistemic linkage between the first-person point of view and third-person study of mind and behavior, and (3) the question of the ontological status of conscious mental events. I will address these issues in turn.

The Role of Phenomenology

For 2,500 years, Buddhist practitioners of mindfulness have been engaged in deep self-reflexive thought. The results of their phenomenological studies are written down, analyzed, revised, nuanced, and taxonomized beginning with the portions of the *Abidhamma* (*Abhidhammattha*, *Abhidharma*) that were written down in the first and second centuries of the Common Era. There is nothing comparable in the West. To be sure, we have Socrates' injunction to "know thyself" followed profitably by the Stoics and Epicureans engaged in sustained attempts to do so. And there is Descartes' insistence on the first-person point of view. But for him it is the importance of the "thin description" of each of us as a "thinking being" that is key, not anything like deep thought about the contents of mind or the interconnections among the multifarious types of mental states. And of course there were, in the eighteenth, nineteenth, and twentieth centuries, phenomenological movements in continental Europe that attempted to probe introspective states but were criticized for so doing by behaviorism and computational cognitive science (in the United States, especially). No matter how many sites of good phenomenology one turns up in the West (the list will, in fact, be long; see Güzeldere 1997), there is nothing remotely like the sustained 2,500-year research program in phenomenological psychology that comes from the Buddhist tradition. (On the Western revival, see Petitot et al. 1999; Noë 2004; Gallagher 2005; Zahari 2005; Revonsuo 2006; Thompson 2007.)

The Dalai Lama writes:

Buddhism and cognitive science take different approaches. Cognitive science addresses this study primarily on the basis of neurobiological structures and the biochemical functions of the brain, while Buddhist investigation of consciousness operates primarily from what could be called a first-person perspective. A dialogue between the two could open up a new way of investigating consciousness. The core approach of Buddhist psychology involves a combination of meditative contemplation, which can be described as a phenomenological inquiry; empirical observation of motivation, as manifested through emotions, thought patterns and behavior, and critical philosophical analysis. (2005, p. 165)

There is no longer any doubt that thick phenomenological descriptions of mental life are important in their own right, as well as necessary for robust theory construction. But there are legitimate worries—one sees the concerns expressed inside Buddhism and they were pressed with a vengeance by behaviorists—about how one knows whether one is seeing mind clearly and describing or analyzing it accurately.[40] We can distinguish three closely related questions of epistemic importance concerning phenomenology: (1) Does phenomenology reveal what mind and its states are like universally? (2) What checks phenomenology other than more phenomenology? (3) Does phenomenology reveal anything more, any thing other, than how mind seems first-personally? Because phenomenology is fashionable again, (1)–(3) require close attention. The good news is that we can make better progress on giving credible answers to (1)–(3) now that we have something to evaluate phenomenological reports against, namely neural activity and behavior.

Linking the Phenomenology with the Brain and Behavior

Buddhist philosophical psychology provides a grand taxonomy of mental states. In part this is due to the effects of deep curiosity and sustained attention on metal life as such. A second reason has to do with the guiding purpose of Buddhist phenomenological inquiry: to contribute to the alleviation of suffering. The parts of the *Abidhamma* devoted to psychology draw distinctions among wholesome, unwholesome, and neutral mental states, and analyze closely the functional links among mental states—for example, what the three poisons (avarice, anger, and false view), and the 24 derivative mental afflictions lead to. (See table 4.2 below.) This elaborate Buddhist psychology is of great interest in its own right, but is meant ultimately to serve Buddhist ethics.

When Buddhist phenomenology developed there was no neuroscience to tether it to. Now there is. We are now positioned to link the first-personal descriptions of mind with third-personal descriptions of behavior and brain. The Dalai Lama writes: "A dialogue between the two could open up a new way of investigating consciousness." This is true. Linking the phenomenological with the psychological and neural is a promising research strategy for understanding persons. It is, furthermore, the strategy now firmly in place (Flanagan 1984, 1991a, 1992, 1996a, 2000).

At times the Dalai Lama protests too much about the lack of recognition of this point in the West. I have been saying this sort of thing for over twenty years in the company of fellow analytic philosophers and scientific naturalists, and even at the start I didn't feel remotely alone. I also know that the late Cisco Varela, a champion of "neurophenomenology," had the Dalai Lama's ear for a long time. One thought I have is that Alan Wallace, who has also had the Dalai Lama's ear for a long time, and who emphasizes more than Varela and I do the comparative lack of attention to first-personal data, residual suspicions about first-personal data, and the relative immaturity of phenomenology in the West, has proven to be the more powerful voice in how the Dalai Lama conceives of the current state of mind science in the West. It isn't that Wallace's description of the situation is wrong. In fact, it is right as an analysis of the overall historical trajectory of Western mind science. But it is not true now—among the best practitioners. So I will assume that we are all in agreement with the Dalai Lama's hope that we collaborate on the long-term heady project of integrating, as far as we can, the first-person and third-person points of view.

That agreed, a wonderful study by Logothestis and Schall (1989) has been widely cited by those of us who favor this blending of the first-person and third-person points of view. Here I treat it as paradigmatic. The experiment is a study in binocular rivalry. When humans are shown vertical and horizontal lines moving across each other they experience Gestalt shifts where from moment to moment they see only the vertical or the horizontal lines. Logothestis and Schall taught, via operant conditioning, rhesus macaques to report (by pressing buttons) when they experience horizontal or vertical lines. By hooking up electrodes to 66 neurons in the areas that do motion detection in the middle temporal and medial superior temporal areas of the superior temporal sulcus (STS), the experiments show that activation in many neurons is reliably dictated by retinal stimulation (of both sets of lines), but other smaller sets are differentially and reliably linked to monkey's phenomenological reports of seeing only horizontal or vertical lines. I want to say that nowadays all good work in cognitive and affective neuroscience utilizes phenomenological reports. Work by Hannah and Antonio Damasio, Richie Davidson, Joseph LeDoux, Christof Koch, and anyone else I can think of who is interested in consciousness uses the first person and looks for brain correlates.

That said, there are legitimate residual epistemic concerns about first-personal data. First, there is the Wittgensteinian concern: We can divide mental vocabulary in two; there are world-directed concepts and mind-directed ones. 'Red' is world-directed. Parents have, at their disposal, a good system of checks and balances to gain assurance that kids use the word 'red' correctly. 'Love', 'sad', 'happy', and the like are mind-directed or just plain mental. Such states have certain world-stimulus conditions that we believe can and do produce them, but these are multifarious, as are the behavioral manifestations of these states. Parents are in an imperfect position when in comes to getting the kids to use these words in the same way they (think they) do or other kids do. We do not worry that these concepts float completely free; it is just that there is less reason to think they are semantically well behaved across speakers of the language than are world-directed concepts.[41] Second, there is the worry about unconscious states. I assume some truth to the idea of a motivated Freudian unconscious and lots of truth to the idea of an unmotivated cognitive unconscious (unconscious language processing, for example). Even with minimal concession to Freud, most will agree that there are self-serving tendencies to keep everything true about one's mind from appearing to oneself (in addition there are things we experience that we sensibly don't say or share).

The cognitive unconscious causes a different sort of epistemic problem, one with ontological significance. Consider the rhesus macaques that fix alternately on vertical and horizontal lines. We might be inclined to say that the activation of the neurons that are reliably, but differentially active, when they report seeing vertically or horizontally *is* the neural underpinning of the conscious percept. This is a credible inference. But whether the active set exhausts or is all there is to the conscious percept is more obscure. Phenomenology is not positioned to say much of anything useful on this matter. Several possibilities remain open. Perhaps more than the set of neurons that reliably tag "seeing vertically" constitute the percept; perhaps fewer are responsible for the phenomenology. At present no one knows, even in the case in question.

Last, there is the Rylean concern, which has to do with making any ontological inferences about the way mind *is* from the way mental states *seem*. The most worrisome inference is immaterialist: Mind, *res cogitans*, seems unextended, immaterial, so it is. Buddhist phenomenology like pretty much all phenomenology across human space and time sees the way mind *seems* as congenial to metaphysical immaterialism.[42]

The Ontology of Mental States

There is anxiety among Western philosophers and scientists about the continued grip of neo-Cartesian views of mind. No contemporary philosopher is a substance dualist, but property dualism charms a few heavyweights. Furthermore, according to polls most ordinary folk believe in something like what Ryle in 1950 called "the official doctrine." Mind or at least mental properties constitute a *sui generis* ontological kind. What kind? Immaterial. Immaterial, but capable of making energy transfers.

Among philosophically inclined mind scientists there are a range of attitudes about how we ought to *naturalize* subjective experience. Why would one care about naturalizing consciousness? The reason is simple. If consciousness can be sensibly thought of as part of the natural fabric of the universe, then we are able to avoid positing ontologically queer substances or properties, and in addition the science of mind can proceed with the hope that understanding mind and mental causation is possible.

Nearly all mind scientists think there are neural correlates for each and every conscious mental event (and for unconscious mental events too). For present purposes, I use Cristof Koch's book *The Quest for Consciousness: A Neurobiological Approach* (2004) as the best statement of the state of the art of consciousness studies.[43] Koch writes: "*There must be an explicit correspondence between any mental event and its neuronal correlates*. Another way of stating this is that any change in a subjective experiential state must be associated with a change in a neuronal state. Note that the converse need not necessarily be true, two different neuronal states of the brain may be mentally indistinguishable." (p. 17) The question of relations between the two levels will be familiar to philosophers as the problem of the supervenience relation (Kim 1993, 2005, 2006).

Throughout his 2005 book the Dalai Lama seems content with the idea that there are neural correlates for every mental state. But he rightly sees that correlations do not constitute identities. There is a perfect correlation between be a living person and having a beating heart. But a living person isn't a beating heart, nor vice-versa. However, even the minimalist neuronal-correlates-of-consciousness (NCC) view is not always advocated by the Dalai Lama. A good example surfaces in this quote by the Dalai Lama in an article published in the *New Scientist* in May 2003. The article is called "On the luminosity of being." Under his given name, Tenzin Gyatso, the Dalai Lama writes:

> Now I'd like to say more about the fundamental nature of the mind. There is no reason to believe that the innate mind, the very essential luminous nature of awareness, has neural correlates, because it is not physical, not contingent upon the brain. So while I agree with neuroscience that gross mental events correlate with brain activity, I also feel that on a more subtle level of consciousness, brain and mind are two separate entities.

This rare statement is important. Usually the Dalai Lama accepts the NCC view, at least the minimalist version of the NCC view: every mental event has a neural correlate. But here he expresses belief in a very strong form of ontological dualism. At least for "the very essential luminous nature of awareness" there are no neuronal correlates of consciousness, because this part of mind is *not* physical. My interpretation for why the Dalai Lama thinks that at least one part of the conscious mind has no neural correlates has to do with a set of ancient beliefs internal to Buddhism. The idea that is implicit in the argument goes something like this:

(i) If real purification is possible, if achieving Buddha-nature is really in the cards for humans, and not simply a perfectionist goal, then achieving it will require realizing a "pure potential" we already have.

(ii) This pure potential consists in realizing the part of the mind that is defiled in *no* way by the three poisons of acquisitiveness, anger, and delusion.

(iii) This part cannot in principle ever have had commerce with anything material, such as the brain, otherwise it might have been defiled and thus lack the required potential.

(iv) Realizing Buddha-nature is possible, and thus we possess necessarily a part of mind that is pure, that cannot in principle be defiled.

(v) What part is that? *Pure luminous consciousness*, a part of mind unsullied by the three poisons, but also that will leave no traces on even

> the most sensitive devices that might be ever created for detecting neural correlates.

One can believe this sort of thing. But I think one needs to use the caveat (discussed above) to do so in the face of the evidence as it now stands. Suppose (what is very controversial) that all parties accept that phenomenologically speaking there is such a thing as "pure luminous consciousness." What is the deep structure of "pure luminous consciousness"? The scientists will say that there is no evidence that "pure luminous consciousness" is immaterial and that it is an inference to the best explanation that it is realized in the brain. But the Buddhist armed with the caveat can truly say science has not proven in a demonstrative matter that pure luminous consciousness is realized in the brain. So pure luminous consciousness is as it seems. How is that? Immaterial and in addition lacking altogether in neural correlates.

On some other occasion I will write more about the epistemic status of the caveat. For now I will simply say this: the caveat permits a Buddhist or anyone else to believe pretty much whatever they want especially if the demand is that there is disproof, where disproof means something demonstrative. If the caveat required concessions when there are good nondemonstrative, that is, inductive, statistical, and probabilistic reasons to give up a belief, then many more concessions of cherished beliefs might be required. This point of course does not apply uniquely to Buddhism, it is a general consequence of taking the growth of knowledge seriously and of being epistemically responsible as knowledge changes.

Most mind scientists, as I have said, believe that there are neural correlates for every mental event (token neurophysicalism). Recall the distinction between token neurophysicalism and type neurophysicalism from chapter 1: Token neurophysicalism says that all conscious mental states are realized in the brain inside a particular body, but token neurophysicalism allows that they might be realized in a wildly disjunctive way. Compare this to the situation of defining 'chair': All chairs are realized, but a chair can be beanbag, with three or four legs, with or without a back, made of plastic, wood, or leather, and so on. Thus, we define 'chair' functionally as "something designed to sit on." Type neurophysicalism says that some set of conscious mental events (say, visual perception), in a specific species at least, will have enough physically in common across instances to classify them of the same type, kind, or class. So compare digestion: There exists a unified account of how digestion works in humans, even though at a fine-grained level there are differences in the biochemistry of individual human digesters (different levels and ratios of the stuff that breaks down food, different metabolic rates, and so on) and in the anatomy of individual digesters (stomachs of different sizes, colons of different lengths and shapes). Most cognitive neuroscientists are looking for a level at which consciousness, conceived as a psycho-biological phenomenon, turns out to be like digestion, where at some level of grain a type neurophysicalist account can be given. Better, across different kinds of consciousness (visual, olfactory, emotional, and so on) such scientists seek unified features that bind that type of consciousness. If no such types are found, the cognitive neuroscientists will rightly be frustrated. But philosophical naturalism is still in play so long as token neurophysicalism is true. Cristof Koch speaks for the majority of mind scientists who seek to vinciate type neurophysicalism when he frames his project as follows:

> The working hypothesis of [my] book is that consciousness emerges from neuronal features of the brain. Understanding the material basis of consciousness is unlikely to require any exotic new physics, but rather a much deeper appreciation of how highly interconnected networks of a large number of heterogeneous neurons work. (2004, p. 10)

Koch explains his strategy at some length:

> Francis [Crick] and I are bent on discovering the *neuronal correlates of consciousness* (NCC). Whenever information is represented in the NCC you are conscious of it. The goal is to discover *the minimal set of neuronal events and mechanisms jointly sufficient for a specific conscious percept*. The NCC involve the firing activity of neurons in the forebrain.... By firing activity I mean the sequences of pulses, about a tenth of a volt in amplitude and 0.5–1 msec in duration, that neurons emit when they are excited. These binary *spikes* or *action potentials* can be treated as the principal output of forebrain neurons. Stimulating the relevant cells with some yet-to-be-invented technology that replicates their exact spiking pattern should trigger the same percept as using natural images, sounds, or smells.

After introducing this idea that he is interested in "the minimal set of neuronal events and mechanisms jointly sufficient for a specific conscious percept" and indicating that the relevant correlates might plausibly be expected to be found in electrical activity, Koch points out that electricity plus biochemistry may well be necessary:

> It is possible that the NCC are not expressed in the spiking activity of some neurons but, perhaps, in the concentration of free, intracellular calcium ions in the postsynaptic dendrites of their target cells. [T]he proposition that the NCC are closely related to subcellular processes is not as outlandish as it may sound. Cellular biophysicists have realized over the past years that the distribution of calcium ions within neurons represents a crucial variable for processing and storing information. Calcium ions enter spines and dendrites through voltage-gated channels. This, along with their diffusion, buffering, and release from intracellular stores, leads to rapid local modulations of the calcium concentration. The concentration of calcium can, in turn, influence the membrane potential (via calcium-dependent membrane conductances) and—by binding to buffers and enzymes-turn on or off intracellular signaling pathways that initiate plasticity and form the basis of learning. The dynamics of calcium in thick dendrites and cell bodies spans the right time scale (on the order of hundreds of milliseconds) for perception. Indeed, it has been established experimentally in the cricket that the concentration of free, intracellular calcium in the omega interneuron correlates well with the degree of auditory masking, a time-dependent modulation of auditory sensitivity in these animals. (Koch 2004, pp. 16, 17)

The fact that Koch is hunting in these two spaces, one electrical and the other biochemical, is important. It means that we need remember that even the best contemporary scientific work does not yet show how conscious percepts are realized, and that we don't yet know which NCCs realize consciousness as a general phenomenon or as a set of phenomena that yield "experience" in different domains (Metzinger 2000).

Epiphenomenalism is the thesis that consciousness has no physical effects. According to the laws of physics (e.g., conservation laws), consciousness, if it is immaterial, can have no effects. None. If a Buddhist insists on operating with the assumption that the mind is immaterial and causally efficacious, his position is scientifically and philosophically unstable. The neurophysicalist will claim that the thesis that mind is not embodied undermines the very idea that "mind-training," conceived non-naturalistically, could yield any of the promised changes in the life of an embodied Earthly being. That said, there is a form of the "epiphenomenalist suspicion" afoot inside current neuroscience. It does not, however, have anything to do with the idea that consciousness might be realized immaterialistically, and it has everything to do with the question of whether we overrate the power of conscious experience in our lives. The emerging consensus is that we probably do (Pockett, Banks, and Gallagher 2006). Freud, of course, said the same thing for different reasons.

Like me, Koch thinks that the weight of all the science taken together require as an inference to the best explanation that (a) consciousness is not epiphenomenal; (b) that eventually "a theory that bridges the explanatory gap, that explains why activity in a subset of neurons is the basis of (or, perhaps, is identical to) some particular feeling, is required" (2004, pp. 18, 19). One reason to hope that the explanatory gap be closed is because if it is not we are left (unless science comes up with radically new laws) with no way to explain mental causation and thus to keep consciousness from, in fact, being epiphenomenal.[44]

In any case, in the most recent iterations of the dialogue between Buddhism and mind science, with the one notable exception concerning "pure luminous consciousness," the Dalai Lama seems comfortable with entertaining the minimalist NCC view. But he then reports an unfortunate exchange with a Western mind scientist about mental causation:

> I vividly remember a discussion I had with some eminent neuroscientists at an American medical school.... I said to one of the scientists: "It seems very evident that due to changes in the chemical processes of the brain, many of our subjective experiences like perception and sensation occur. Can one envision the reversal of this causal process? Can one postulate that pure thought itself could effect a change in the chemical processes of the brain?" I was asking whether conceptually at least, we could allow for the possibility of both upward and downward causation.... The scientist's response was quite surprising. He said that since all mental states arise from physical states, it is not possible for downward causation to occur. Although, out of politeness, I did not respond at the time, I thought then and still think that there is as yet no scientific basis for such a categorical claim. The view that all mental processes are necessarily physical processes is a metaphysical assumption, not a scientific fact.... In the spirit of scientific inquiry, it is critical that we allow the question to remain open, and not conflate our assumptions with empirical fact. (2005, pp. 127–128)

To be kind to the American scientist, his answer was acceptable *if* he understood the question to be can mind science make any sense of the idea that non-physical events (i.e., events that have no matter, have no energy, and contain no information) can affect anything? There is nothing in any science, at this time, that allows for that kind of causation. And thus if the mind has no physical properties, it does no causal work. Consciousness is epiphenomenal.

But let us interpret the question as follows: Is there any problem in contemporary mind science with the idea of transformation of mind by mind?

The answer is No, but only so long as mental states have causal powers. Everyone who believes that transformative practices work ought to believe (I claim) that they work because of the way in which some mental (sub)systems can gain control over other (sub)systems. Consider the universal experience of being tempted to do something you consider wrong, recognizing it would be wrong, and not doing it. The mental, phenomenological feel involved in such a process reveals no neural texture, but according to the ontological commitments of our best science we assume it is in fact embodied. What we now know but didn't even fifteen years ago is that high cortical areas overrule lower ones in such cases (LeDoux 1996). Some scholars think that the efficacy of mind training practices (meditation, mindfulness-based stress reduction, etc.) vindicate the Dalai Lama's view that mind conceived non-physically can affect the brain and body (Begley 2007; Wallace 2007). While it is true as Sharon Begley and Alan Wallace report (and as I did in 2000a, 2003a, 2006a, and in chapter 5 below) that there is serious and much welcomed ongoing collaboration between Buddhists and neuroscientists on the way certain Buddhist practices take advantage of brain plasticity, it is false that this vindicates the philosophical idea that immaterial mental events can or do cause the changes in the subjects. The view that immaterial stuff or properties can cause anything to happen is an idea, which despite numerous noble efforts no one has ever been able to make sense of. The idea that it is possible is inconsistent with all known scientific law. "From something immaterial *nihil fit.*"

Furthermore, the naturalist has a credible view on offer: we have evolved as animals who are highly sensitive to certain information and we have remarkable reflexive capacities. As in the case of prefrontal cortical control over lower brain regions in cases where we overcome temptations, so too cases of mind-training are best explained as follows: The conscious experiences of, say, a meditator are realized in his brain, and they either cause some brain systems to utilize pre-existing control capacities over other brain areas, so that, for example, he can attend quietly to his breathing alone, or, as likely, a training effect occurs where practice adjusts the interaction among various brain systems to yield, say, greater patience. No immaterial forces need be imported.[45]

What about doing as the Dalai Lama recommends and avoiding the deep philosophical issues for now? My view is that we should distinguish between two questions, only one of which we should now leave open.

There is the question of whether mental states might be immaterial. Then there is the question of how the embodiment of mind actually works, how exactly mental states are realized in the brain, why such and such neural activity produces blue experiences rather than red experiences, and so on. Leaving the first question open requires that we accept that we are clueless about how mental causation is possible, and, what is different, what ingredients are necessary for it. We aren't clueless about these matters. It is an inference to the best explanation that our world is a natural one, that consciousness is realized in the brain, and that mind has the causal powers it has because it is so situated.

Leaving the second question open is simply judicious, since we don't know its answer. If we accept neurophysicalism about mind as the right way to conceive answering the first question, then at least we know where to look for an answer to the second question.

The Dalai Lama thinks it wise to keep both questions open. He writes this just after reporting his conversation with the American scientist:

> It may well be that the question of whether consciousness can ultimately be reduced to physical processes, or whether our subjective experiences are non-material features of the world, will remain a matter of philosophical choice. The key issue here is to bracket out the metaphysical questions about mind and matter, and to explore together how to understand scientifically the various modalities of the mind. I believe it is possible for Buddhism and modern science to engage in collaborative research in the understanding of consciousness while leaving aside the philosophical question of whether consciousness is ultimately physical. (2005, pp. 136–137)

This can be done, that is, we can "bracket out the metaphysical questions about mind and matter . . . leaving aside the philosophical question of whether consciousness is ultimately physical." But if we do so in the spirit of thinking that mental events might be or turn out to be nonphysical and thus possessed of no causal powers, then we are being insincere. If all we were now doing in mind science was a mapping between first-personal experience and the brain, then maybe we could do the bracketing in good faith. But mind science is already much more advanced that that. We are now doing this sort of mapping while at the same time trying to figure out the causal relations among various components of mind and the relations among mind, brain, and behavior. And there is a vast amount of research now about how higher cortical regions control lower brain regions, as well as the other way around. So neutrality on the metaphysics of mind is not, as James would say, a live option.

What is a live option is this: Keep an open mind about how conscious mental states are realized "neurally" while assuming that they are. Once there was a view that there would be neat one-to-one mappings between the phenomenal and the physical (along the lines of such perfect identities as "Water is H_2O"). There is still some hope for something like identity theory for sensations. But almost no one believes that strict identity theory will work for more complex mental states. Strict identity theory, as I use the term, is stronger that type neurophysicalism. Strict identity theory would involve exhaustive, reductive identities between some conscious mental state type and a type of brain process at the level of biochemistry. The analogy with water is a good one: The stuff we call 'water' as described at the macro-level is identical with the molecular stuff we call 'H_2O' at the lowest micro-level. But no one committed to type neurophysicalism believes that type-identities will be strict all the way down to the lowest levels of neural grain or texture. That said, token neurophysicalism will obtain all the way down.

The judicious strategy is to wait and see how the mapping goes. It is likely to be very complex, with bridge principles that will need to be invented. And come what may, no nasty reductionist or materialist will be in any position to say that consciousness is an illusion or that you don't make choices. That said, the best hypothesis is that the conscious mind is the most complicated biological phenomena ever studied. It is precious and beautiful and it is part of the natural fabric of the universe.

It is my sincerest hope that the already immensely profitably dialogue between Buddhism and science continues to develop, and that we engage each other as honestly as possible in a spirit of mutual respect. That is what I have tried to do here.

Appendix 1: Buddhism and Rebirth

History

In his magisterial book *Imagining Karma: Ethical Transformation in Amerindian, Buddhist, and Greek Rebirth* (2002), Gananath Obeyesekere distinguishes between two kinds of rebirth theories: non-karmic eschatologies and karmic eschatologies. The first are most common in small-scale societies. There are two typical belief: a dead member of the community will circle back into the community; a newborn can be identified as the reincarnation of a specific ancestor.

We can understand the appeal of the idea. Grandmother has died and is missed; her grandchildren are bearing children. A newborn daughter is seen as grandmother's reincarnation. (In a small society there will be, for now familiar Darwinian reasons, strong physical "family resemblances.") One sees the idea of non-karmic rebirth with frequency in Amerindian and West African tribes. There are exceptions in the ethnographic literature, e.g. Trobrianders.

In larger societies with greater mobility, especially if community members exit or new ones enter, one sees the emergence of karmic eschatology: There is rebirth. But it may occur far away. Sometimes this involves going to heaven or hell for eternity, which involves going very far away (usually to be joined later by one's loved ones). Or it may involve entering a cycle of rebirths, until—according to most schemes—some form of final liberation/release/dissolution/heavenly life is achieved. The key to the "quality" of one's next life is one's good or bad karma in one's current life (adding in perhaps karmic effects of previous ones). Rebirth is moralized in this way.

Sticking with the Indic religions one finds both types of karmic eschatology. Obeyesekere writes: "Consider the *Rg Veda*, the oldest stratum of Brahamanic religion [now dated at circa 1500 B.C.E.]. The soul at death, driven by a chariot or on wings, takes the route of the Fathers and reaches a place of eternal rest. The *Rg Vedic* notion of heaven is a paradisiacal one; "there is light, the sun for the highest waters, every form of happiness." There is music, drink, and merriment.

At the time of the *Rg Veda*, Indic religion had operated on a "one life and then heaven" premise. "Rebirth" happens exactly once: when your life on earth is over. Therefore, in 1500 B.C.E. the ideas of many reincarnations and rebirths, innumerable lives to reach liberation, were not in play.

Beyond the *Rg Veda*, "the association between karma and rebirth is not at all clear in the earliest texts and discourses on Indic religions. There are virtually no references to rebirth or to an ethical notion of karma in the Vedas or in the Brahmanas, the oldest texts belonging to the Hindu tradition. The first significant references to multiple reincarnations and rebirths appear in the early Upanishad, the *Brhadaranyaka Upanishad*, probably composed sometime before the sixth century B.C.E., followed by the *Chandogya* and the *Kausitaki*. A hundred years or more later these theories appear in full bloom in the so-called heterodox religions—particularly in Buddhism and Jainism—that have karma and rebirth at the center of their eschatological thinking. Soon afterward these ideas surface in mainstream Hinduism itself and become an intrinsic part of the eschatological premises of virtually all Indic religions." (2002, p. 1)

I get the impulse behind karmic eschatologies. Had I been in charge of the choice between the one life and you go to heaven of the *Rg Veda* option and the later karmic cycle of rebirths option, I would have gone for the former. For one thing, it is much quicker. For another, there is only paradise.[46]

From what Obeyesekere says, there is reason to think that no choice was made, since the *Rg Vedic* idea was never really picked up in the first place. In fact, he thinks the basic idea was not really ethicized because there was only the paradisiacal outcome, not a punitive one. A workable karmic eschatology requires a sensible system of rewards and punishments. Thus, there was a vacuum and the idea of the karmic rebirth cycle filled it.

Thus, as we have seen, in India the idea appears no later than 600 B.C.E.—some accounts think it appears as early as 900 B.C.E.—and it spreads in the East like wildfire (Similar ideas appear in the West in Pythagoras and Plato and eventually in all three Abrahamic religions, in the latter cases in something akin to the *Rg Vedic* idea but ethicized thanks to "hell"). The key feature of all forms of karmic eschatologies is that they moralize life in a specific metaphysical way. There is the normal earthy system of payoffs for living well. But there is, in addition, a metaphysical system of payoffs that operates after death. In orthodox Indic traditions, my identity, my essence is constituted by my *atman*, my soul. It is not 100 percent clear what or who orchestrates the system of karmic payoffs although what happens is pretty straightforward. Depending on the quality of this life, one's *atman* is reincarnated into a better or worse, higher or lower, human or non-human life. Perhaps Brahman, the creative source behind the universe, "makes the decisions" about reincarnations or, more likely, because Brahman is pretty impersonal by Abrahamic standards, the laws of karma work somehow up and alongside ordinary causal laws. How this came to happen is obscure.

My Analyses and Commentary

Now Buddhism is distinctive in several ways. It is heterodox. First, It is not theistic. It is either atheistic or quietistic on theological matters. Second, there is the denial that I am, or have, an immutable *atman*. A person is a psychologically continuous and psychologically connected being. But personhood is part of the flux. Without going into the various interpretations of the *anatman* doctrine, we can see why, if Buddhists are going to talk about future lives, most careful ones will distinguish between reincarnation (in which my *atman* is reborn) and rebirth (the view that—maybe—the-conscious-stream-of-being-that-I-am continues on in some way). Think of the idea along these lines: a naturalist might think, using conservation principles, that when I die the stuff I am made of disperses and rejoins the universe. If I am *anatman* I might believe that the consciousness that constitutes me is (a) immaterial and (b) resurfaces in another living being in a way suited for *anatman* rather than an immutable, indestructible *atman*. In my experience many Buddhists have serious difficulty explaining how the continuity works and in what way "my" consciousness continues. One teacher I have sat with at the Kadampa Center in Raleigh North Carolina, Ven. Robina Curtin, an ordained Australian in the Gelug lineage, well known for her film *Chasing Buddha*, always says that Buddhism falls apart unless you believe that mind is immaterial and also that somewhere between 5 and 7 weeks before a human impregnation event, the next "soul" in-waiting is in the vicinity of where the conception event occurs! In addition to describing such an improbable event with such precision (how is the time frame known?), she reverts to describing the rebirth in a way that more suits reincarnation. That is, she—and in my experience—many wise and learned Buddhists, have trouble describing how the continuity of *anatman* works without using the rejected conceptual categories appropriate to atman. It is I, Owen the *atman*, that gets reborn, perhaps in the body of a sewer rat or a bodhisattva, but it is I. Indeed, in Tibetan Buddhism one sometimes hears that enlightened beings, Dalai Lamas, for example, get to plan and/or orchestrate their next rebirth. Actually, this doctrine is easy to find in the esoteric tantric teachings. I can in fact see the psychological and motivational force that playing with such an idea might have for someone who is trying to do what is immensely hard and worthy: be as excellent as you can be. But I can see no reason to think it true.

In the very same inaugural address in which the historical Buddha expressed dismay over the question of God's existence, he claimed to be clueless about whether he (or anyone else) had any sort of afterlife. Of course, the fact is that there is abundant evidence in other sutras of the Pali canon that the Buddha endorsed the belief in rebirth, even remembering aspects of former lives. And, on the ground, among ordinary Tibetan Buddhists even if there is no god, there are lots and lots of spirits, heaven and hell realms, and widespread belief in rebirth (often of the Hindu "reincarnation" sort). In order to be brief I will get right to my best guess diagnosis about why the belief is so solid, resilient, and almost universally avowed among Tibetan Buddhists: Buddhism, like Jainism, *is* heterodox relative to "orthodox" Hinduism. But much is absorbed and maintained from the Mother Indic roots of all three (including Hindu) traditions. Obeyesekere points out that what one finds in the roots of what was to become "orthodox" Hinduism are two threads: "[I]n [some passages] of the Upanishads (and similar doctrines among both Indian and Pythagorean Greeks) is a preoccupation with the individual's personal salvation.... By contrast, in the samanic religions, such as Buddhism and Jainism, there was an additional interest in the soteriological-cum-ethical welfare of common men, women, and Sudras, that is, of the larger lay community." (2002, p. 113) So sociological reasons having to do with the "Mother tradition" helped the doctrine along, especially perhaps in the cases of Buddhism and Jainism a special concern with communal ethical perfection over individual salvation. Then, and I don't think this can be underestimated, there is this motivational problem faced by every great spiritual traditions, especially one, such as Buddhism, with perfectionistic aspirations. If I only have one life, then my prospects for achieving enlightenment in this very life do not seem that likely (although see Pandita 1992 for an exploration of the "in this life" possibilities from a Theravadan perspective). In my 1991 book *Varieties of Moral Personality: Ethics and Psychological Realism*, I argued that in ethical and spiritual life as in athletics, when traditions set goals unrealistically high, there is often a loss of motivational hold on advocates. So by developing the idea that I am responsible for many past and future lives, I may see that I do in fact have "enough" time to become wise and virtuous; and, in addition this may make me as motivated as is possible to try very, very hard. I see this idea in some of Robert Thurman's writings.

There are in the West some Buddhists who think that both the theistic impulse and unwarranted ideas about karmic laws or karmic eschatology resurface much too often in Buddhism and should (gently) go away. The reasons are pretty straightforward: such ideas are conceptually unstable without a God to orchestrate the karmic system and without a substantial soul to be continuously re-embodied. Never having heard any good arguments for rebirth that do not simply appeal to dubious memory sources, or some form of an argument from authority, or invalid logical arguments, I share this opinion. But nothing follows from my opinion about what especially non-Western Buddhists ought to believe. I am not sufficiently inside the practice to understand how it works. Buddhism is a noble wisdom tradition with its own internal standards from which it generates its beauty and moral majesty. I think the belief in rebirth is irrational, but many beliefs (see the discussion of "positive illusions" in chapter 5) are irrational and do no harm, only good. The belief in rebirth has, it seems to me, this sort of innocence. So from where I sit I leave it alone unless it is brought into a discussion of mind science. Then and only then does my opinion count.

That said, I do think that the conceptual adjustments called for by Stephen Batchelor (1997) and other Western Buddhists may well become dominant in Western Buddhism. First, there is simply the fact that Buddhism, as I write, is especially attractive to Westerners who are tired of Abrahamic theism. Second, many think of themselves as secular naturalists. Third, among philosophers who work on such issues as the nature of the mind, personal identity, and ethics, the doctrine of *anatman* combined with the social interpretation of the doctrine of co-dependent origination (as some Africans say "I am because we are," combined with the importance of becoming detached from such things as "personal salvation," make Buddhism especially attractive while, at the same time, making the tradition particularly poorly suited for the doctrine of karmic rebirth. What is beautiful about Buddhism can perhaps be retained in a (more) conceptually sophisticated and stable form without the doctrine of rebirth.[47]

Appendix 2: Catholicism and Evolution: Can a Roman Catholic Be a Darwinian?

Caveat: Papal Infallibility

There is a lot of confusion on this matter. In *strictu sensu*, a papal pronouncement is only infallible when it (a) pertains to a matter of faith or dogma and (b) is pronounced according to the very formal rules governing speaking *ex cathedra*. I believe that the last time (a) and (b) happened was in the 1950s, when the Blessed Virgin's direct bodily assumption into heaven was made a formally required, infallibly asserted piece of dogma. So some things like the Roman Catholic views on contraception have not been made infallibly, but perhaps they follow deductively from some doctrine that has infallible status. This issue of logical implication causes an interpretive complication about the matter of evolution. Although none of the statements that pertain to evolution were stated in the infallible *ex cathedra* mode, they do (I am pretty certain) logically follow from matters of faith and dogma that were so stated long ago with the imprimatur of infallibility.

My friend Michael Ruse has written an interesting book called *Can a Darwinian Be a Christian?* (2001). His answer is "Yes, but it is not all that easy." Here I turn the question around and ask more specifically "Can a Roman Catholic be a Darwinian?" (or, better perhaps, *should* a Roman Catholic be a Darwinian?").[48] The force of 'should' here is intended to put us in the vicinity of this more specific question: If you believe what Roman Catholicism says you must about God and the soul can you sensibly, coherently believe in evolution? I think the answer is No—if the Catholic accepts what the Church requires. Strict Catholic teaching will cause the "wannabe" Catholic Darwinian to hold strange, possibly incoherent views on divine intervention, immaterial mental entities, as well as unstable or incoherent views on mental causation. I will offer my reasons for thinking this after I give a reprise of the Church's position on evolution. If I am right, it helps us to understand why (in America and perhaps elsewhere) the numbers who do not accept Darwinism are so high—in the 50 percent range. Besides those who do not understand what Darwinism in the question is, most religions are not really logically compatible with Darwinism.

I. In his 1950 encyclical *Humani Generis,* Pope Pius XII made some conciliatory gestures toward those who wish to study the development of embodied life on Earth. However, it is important that the encyclical bears this subtitle: CONCERNING SOME FALSE OPINIONS THREATENING TO UNDERMINE THE FOUNDATIONS OF CATHOLIC DOCTRINE TO OUR VENERABLE BRETHREN, PATRIARCHS, PRIMATES, ARCHBISHOPS, BISHIOPS, AND OTHER LOCAL ORDINARIES ENJOYING PEACE AND COMMUNION WITH THE HOLY SEE

The contents pertaining to evolution can be summed up as follows:

1. "The Church does not forbid that ... research and discussions, on the part of men experienced in both fields, take place with regard to the doctrine of evolution, in as far as it inquires into the *origin of the human body* as coming from pre-existent and living matter." (emphasis added)

2. Catholics are free to form their own opinions on the evolution of the body, but they should do so cautiously; they should not confuse fact with conjecture, and they should respect the Church's right to define matters touching on revelation and ethics.

3. Catholics must believe, however, that the human soul was created immediately by God. Since the soul is a spiritual substance it is not brought into being through transformation of matter, but directly by God, whence the special uniqueness of each person.

4. All men have descended from an individual, Adam, who has transmitted original sin to all mankind. Catholics may not, therefore, believe in "polygenism," the scientific hypothesis that mankind descended from a group of original humans (that there were many Adams and Eves).

Some have called the pope's view neutral. But no (non-Catholic) neo-Darwinian will accept 3. And many will have trouble with 4. Others describe the view as "compatibilist," as an attempt to keep the two theories from contradicting each other. After I say more about the history of the Church's views, I will explain why I think no real compatibility is attained.

II. In 1996, Pope John Paul II moved beyond Pius XII's judicious openness about the possibility of natural evolution of the human body to acceptance of the story as true about the body. John Paul II wrote:

> In his encyclical *Humani Generis* (1950), my predecessor Pius XII has already affirmed that there is no conflict between evolution and the doctrine of the faith regarding man and his vocation, provided that we do not lose sight of certain fixed points.... Today, more than a half-century after the appearance of that encyclical, some new findings lead us toward the recognition of evolution as more than an hypothesis. In fact it is remarkable that this theory has had progressively greater influence on the

spirit of researchers, following a series of discoveries in different scholarly disciplines. The convergence in the results of these independent studies—which was neither planned nor sought—constitutes in itself a significant argument in favor of the theory.

In the same "Message to the Pontifical Academy of Sciences on Evolution," John Paul II reinterated the rejection of any theory of evolution that provides a materialistic explanation for the human soul:

Theories of evolution which, because of the philosophies which inspire them, regard the spirit either as emerging from the forces of living matter, or as a simple epiphenomenon of that matter, are incompatible with the truth about man.

III. In part because a new pope, Benedict XVI, was installed in 2005, at a time of continuing and sometimes ferocious debates about creationism and intelligent design, people have had their ears open to any new pronouncements by the Vatican. A considerable stir was caused by a *New York Times* op-ed piece in July 2005, "Finding Design in Nature," in which Cardinal Schondorn, a close colleague of the new pope and the cardinal archbishop of Vienna, wrote:

Ever since 1996, when Pope John Paul II said that evolution (a term he did not define) was "more than just a hypothesis," defenders of neo-Darwinian dogma have often invoked the supposed acceptance—or at least acquiescence—of the Roman Catholic Church when they defend their theory as somehow compatible with Christian faith.... But this is not true. The Catholic Church, while leaving to science many details about the history of life on earth, proclaims that by the light of reason the human intellect can readily and clearly discern purpose and design in the natural world, including the world of living things.

Evolution in the sense of common ancestry might be true, but evolution in the neo-Darwinian sense—an unguided, unplanned process of random variation and natural selection—is not. Any system of thought that denies or seeks to explain away the overwhelming evidence for design in biology is ideology, not science.

In an unfortunate new twist on this old controversy, neo-Darwinists recently have sought to portray our new pope, Benedict XVI, as a satisfied evolutionist. They have quoted a sentence about common ancestry from a 2004 document of the International Theological Commission, pointed out that Benedict was at the time head of the commission, and concluded that the Catholic Church has no problem with the notion of "evolution" as used by mainstream biologists—that is, synonymous with neo-Darwinism.... The commission's document, however, reaffirms the perennial teaching of the Catholic Church about the reality of design in nature. Commenting on the widespread abuse of John Paul's 1996 letter on evolution, the commission cautions that "the letter cannot be read as a blanket approbation of all theories of evolution, including those of a neo-Darwinian provenance which explicitly deny to divine providence any truly causal role in the development of life in the universe."

Furthermore, according to the commission, "An unguided evolutionary process—one that falls outside the bounds of divine providence—simply cannot exist." . . . Indeed, in the homily at his installation just a few weeks ago, Benedict proclaimed: "We are not some casual and meaningless product of evolution. Each of us is the result of a thought of God. Each of us is willed, each of us is loved, each of us is necessary.

My Analysis

1. All parties above agree that God is creator. This is what is means to be a theist. So if natural selection explains bodily form/evolution, God being all-knowing and all-powerful planned it that way. He might have laid down the plan when he planted the singularity that banged and created the cosmos, or he might have intervened somewhere along the way during world-historical time and added Darwinian principles to the mix, thinking them a good idea. If the first, we have the God of the deists, which is less that the Roman Catholic Church wants. If the second, then God, a purely spiritual/immaterial being, causally interacts with the world in real time. This is an idea that no one has ever made any sense of. What has happened, of course, is that cosmologists, evolutionists, and so on will politely say that creation *ex nihilo* is not something they can rule out. But truth be told it is not something they can provide any sensible traction to, given an overall commitment to known physical laws. Allowing something like the God of the deists is, as far as I can tell, a matter of political politeness among those who have thought their science through but see no reason to bother those who avow such a (they hope) innocent belief. There are also, alas, many scientists who have simply not thought their science through and thus believe that it is epistemically responsible for them to believe in a creator God. It isn't. Buddhists have the better idea here. Be an atheist, agnostic, or quietist on this question.

2. "Theories of evolution which, because of the philosophies which inspire them, regard the spirit either as emerging from the forces of living matter, or as a simple epiphenomenon of that matter, are incompatible with the truth about man." And thus we must believe that at conception God "implants" a unique personal soul.

Two Comments on Item 2: Soul Implementation at Conception

First, evolution does regard sentience, including human sentience, as an emergent biological phenomenon. But we need to understand why this is a problem beyond the fact that it is inconsistent with the "soul implementation" doctrine. And we are not told this. But, second, it is false that evolution treats "the spirit" as a "simple epiphenomenon of matter." If "spirit" means an immaterial soul, then evolutionary thinkers will not buy into the existence of such a thing. If it means "the embodied conscious mind," then fine. The conscious mind and its causal powers are not a problem, at least not in principle. Each day neuroscience is unlocking new secrets about the way our embodied minds work. Subjectivity is real, it emerged evolutionary and its home is in our brain/embodied selves. An epiphenomenon is a side effect of some process which itself does no interesting causal work. But minds, according to the theory of evolution, do lots of work. We have minds, we perform intentional actions, etc. Our karmic effects are abundant and visible everywhere. Furthermore, each mind is unique because it emerges from a unique set of biological and developmental processes. If God implants souls, then he is exerting the sort of causal powers in real-time that no respectable scientist should think possible. The situation is much worse even than the intervention I imagined when God decided to bring natural selection to the world—say, when the big bang banged, or when Earth came to be. He is literally involved in each and every human impregnation event!

3. There is another problem. Either souls do causal work or they don't. If they do causal work, they don't do so according to any principles science accepts. If they do no causal work they are epiphenomenal!

4. Why do Catholics care about immaterial souls? My view is that it all comes down to Catholic eschatology. Bodies die, decay, and disperse. Perhaps immaterial stuff doesn't. Thus the idea *seems* to be required if the usual views of eternal reward and punishment have any chance of being credible. If one wants to play this mysterious game, then one ought to be imaginative. There is another obvious route one could go. Let God be the deist one, allow his plan to be for the world to unfold as he planned, positioning us to see that goodness will pay. Have God from outside this world keep score on each life. When you die allow God in His world to create a replica of you in immaterial ectoplasm. He rewards "you" in "His

World" as he sees fit. Some, not all, problems with interaction would become less problematic if theologians took this advice.

Because this idea concedes a bit to the idea of an all-powerful and all-knowing God in some other world, it might be something that scientists will let it pass, not because they think it true, but because they see the beliefs involved as relatively harmless. Like David Sloan Wilson in his book *Darwin's Cathedral* (2003), I think we sometimes overemphasize the importance of incredible theological ideas in religion and lose sight of some of the important moral wisdom they often contain (but admittedly, it's a mixed bag).

The upshot of the discussion is this: If you are a true believer in what the Roman Catholic Church says you *must* believe about God, his continuous intervention in every human impregnation event, and about the immaterial nature of souls, then your commitment to what science teaches is very weak. The theory of evolution is very well confirmed and it denies or entails the denial of what you as a Roman Catholic are required to believe. There is a type of Roman Catholic that can believe in evolution. I perhaps fit this bill: Believe none of the theology or metaphysics. But be a cultural or ethnic Catholic (the way many Jewish atheists are). Go to Mass, meditate and pray in a Catholic way if you wish, consult the right saints depending on your needs, have fun, etc. This is a reasonable way of affirming your identity, you can find wise moral guidance in places, and you can drop all the hocus-pocus stuff. That stuff is silly, unbecoming to thoughtful souls, and can be dangerous.

4 Normative Mind Science? Psychology, Neuroscience, and the Good Life

Can There Be Such a Thing as Normative Mind Science?

The naturalistic picture of persons is not inherently deflating or disenchanting. The reason it is not deflating is that it accepts and thus does not deny that we are conscious and that consciousness gives us some control over how we live. The reason it isn't disenchanting, at least not necessarily, is that our remarkable powers as persons remain. We are creatures who can and sometimes do make sense of things and find meaning. Our nature and our powers are explained, perhaps, but they are not eliminated. Furthermore, the almost unimaginable complexity of naturalistic explanations or explanation sketches of our kind of being reveal the beautiful depth, texture, and intricacy of our being, even if they remove whatever undeserved enchantment comes from mystifying analyses with numerous slots for the variable "and then the miracle occurred."

We are biological beings living in a material world that we have constructed. Our norms and values are designed to serve our purposes as social mammals living in different social worlds. History, and possibly our psychology, has led us to mystify norms and values.

It would be nice if, in addition to showing this much, I could also advance the case for eudaimonics, for the idea of empirical inquiry into the nature, causes, and constituents of flourishing, and in particular if I could advance the idea that some ways of living and being are better than others. Empiricism, it seems to me, is underestimated as an excellent (let alone as the best) source of true wisdom about our nature and our situation. One reason is that empiricism is sometimes painted as a mundane and uninspiring approach to world-making given what might seem to be attractive ways of mystifying our situation. My own view is that an empirically inspired eudaimonics can help cure the disease of disenchantment by marking off reliable ways to flourish from ways of living and being that tend not to make for meaningful lives while at the same time acknowledging the beauty and majesty of what there is, including us. One norm that I avow and claim that eudaimonics favors is that we ought to seek to flourish with the truth by our side. Thinking about our natures and our lives in ways that incorporate superstition and wishful thinking is childish and unbecoming to rational social animals such as us.

Can there be such a thing as normative mind science? And if there can be, is that a good thing? One might answer "Obviously yes" to the possibility question, since when mind scientists tell us how a proper visual system works, or what the right balance of neurotransmitters is, they are utilizing normative concepts such as 'proper functioning' and 'right balance'. True. On the other hand, a scientist who is shy about doing anything that has an odor of "normativity" could take cover in the latter cases by claiming that such normative concepts as 'proper functioning' and 'right balance' are simply shorthand for what is statistically normal. So let us ask a more revealing question: Can there be a mind science (or, better, inquiry) that empirically studies what is statistically abnormal but nonetheless good and of great value—namely, the nature, causes, and constituents of eudaimonia? Yes. In fact, such work is ancient, and it is now on again. In the twentieth century, the great ancient attempts to do eudaimonistic scientia were, I think, for reasons having to do with positivistic conceptions of science, marginalized and treated as "merely philosophical." In part, therefore, my project is one of re-collection and reclamation.

One can imagine a host of worried responses to the question "Would it be a good thing if mind science were to tell us how to achieve eudaimonia?" In the first chapter, when I first talked about science and the scientific image and what gets some people edgy about them, I pointed to attempts by science to act as if it can do more work than it can in fact do, like explain everything. I also indicated that science describes and explains but does not typically trade in oughts. The word 'typically' is important here. There are and always have been sciences, such as medicine and (more recently) psychiatry, clinical psychology, and social work, that incorporate norms in the relevant sense. Do the norms of physical and mental health come from inside medicine, psychiatry, etc., or from outside those sciences? I think the answer is that the norms come from within these sciences *and* from outside them.

Consider norms of mental health. First, there is observation of what is statistically normal, what the average person is like. Second, the image of the "average Joe and Joanne" is considered in light of norms that come from comprehensive views of what it means to be "mentally healthy" and, what is different but overlaps, what it means to be a "good person." These normative conceptions are in the air in public social space, and they do not abide giving the imprimatur of 'healthy' or 'good' to what is statistically normal—unless, that is, one who is normal is also mentally healthy and possibly spiritually fit and morally good.

People commonly have difficulty in the areas of love and work. There are norms about being a good husband, wife, father, mother, or child. These are culturally and socially variable. It may be statistically normal for husbands to be domineering. Perhaps the best ethical, political, and spiritual thought in the culture says this is wrong. The norms that people need to abide are not the "normal" ones. Mental health workers see the distinction and work with patients to reconfigure themselves and their lives to be better persons. Greater tranquility and happiness might be offered as rewards. Thus, in a first pass we can say (somewhat artificially, because it makes the process temporally two-step and sequential when it is in fact dynamic) that scientifically trained mental health workers develop norms of mental well-being that use the mental health equivalents of homeostasis and equilibrium from ordinary medicine and adjust, modify, and enhance them by utilizing the best available cultural wisdom about what makes for a sane, non-neurotic, non-psychotic person—for being a good parent, a good spouse, and so on. If someone says this is highly fallible, I agree. But what method isn't? One attractive feature of the picture, meanwhile, is that all the spaces in the Space of Meaning$^{\text{Early 21st century}}$ have a say in the norms.

It is worth mentioning that psychiatrists and clinical psychologists, on the one hand, and social workers, on the other hand, typically have different clienteles. Psychiatrists and clinical psychologists typically have clients from the high end of the socio-economic ladder for whom the external conditions of life are good; social workers typically deal with people whose psyches are damaged or in jeopardy because their environments are dreadful. Different circumstances, different treatment issues.

The objection I hear in my ear is this: "Owen, the sciences you are talking about are not really sciences, not really 'scientific'—even practitioners will often refer to what they do as a craft or, even worse, an art. My response is "Compare these sciences to structural engineering. The laws of physics, material science, and engineering are background for the design of a skyscraper. Siting is a particularly important initial condition; then there is the weather issues—which ways the winds blow, whether other structures create wind tunnels, and so on. I remember distinctly when I was in graduate school in Boston just after the John Hancock Tower was finished; its windows kept blowing out. Someone hadn't thought through the particulars about the wind."

When it comes to mental health, or to flourishing, each person is a unique spatio-temporal site, with a particular life history, a certain "normal weather within," situated inside a certain culture that pulls for display of its own norms, and so on. It is more complicated than siting and constructing a building. The "artful" or "craft-like" part comes from applying the scientific or empirical knowledge, wisdom, or generalizations in the nuanced way required by the complex particularity of any human life. In any case, nothing at all turns, for my purposes, on how "scientific" one understands eudaimonics to be, so long as one sees that it is thoroughly empirical. (Recall that the word 'scientia', as I use it in association with the study of eudaimonia, is intended to mean empirical knowledge in a general sense.[1])

Two Ancient Examples of Eudaimonistic Scientia

I said that normative mind science is ancient. I offer two examples that come from approximately 2,500 years ago. The relevant texts are Aristotle's *Nicomachean Ethics* and the Buddhist *Abhidhamma*. Both of these ancient texts provide psychologies of the statistical normal. But both Aristotle's *Nicomachean Ethics* and the Buddhist *Abhidhamma* also provide a theory of human excellence. For Aristotle, such a person is eudaimon—one who is truly happy, one who genuinely flourishes, one who is rational and virtuous. In the Buddhist case we are provided with divisions among wholesome, unwholesome, and neutral mental states, as well as various lists of afflictions and virtues. The most excellent person is the one who has released himself from various poisons and mental afflictions and who lives among the four divine illimitables: compassion, loving-kindness, sympathetic joy, and equanimity. And, like Aristotle's eudaimon, this person—think of him as a *bodhisattva,* a Buddhist saint—is statistically abnormal but embodies the most excellent norms. Both flourish in ways normal folk don't. How does one do normative science of this sort, or of these two sorts? I claim that one does such normative mind science, when one does it well, empirically.

The Aristotelian picture is familiar; I have used aspects of it already in previous chapters, so I will be brief: Everyone agrees that the greatest good is eudaimonia, "happiness." But there is disagreement about what brings happiness or, what is different, constitutes happiness. Every idea is a contender—sensual hedonism, solitude, great wealth, power, and so on. Indeed, all these contenders are live options today. None are the answer. The answer is that our human end—the way we actualize our full potential—is by living a life of Reason and Virtue. Virtue consists of the set of dispositions to perceive, feel, think, judge, and act in the right way, at the right time as the particulars of the situation that calls for the virtue in question warrant. Aristotle's list of virtues includes four from Plato (Courage, Justice, Temperance, and Wisdom), to which six are added: Generosity, Wit, Friendliness, Truthfulness, Magnificence (spending lavishly of worthy things, e.g., sacrifices, warships, public buildings), and Greatness of Soul (believing because it is true that one can accomplish great things and thus is worthy of honor). How did Aristotle—for that matter, how does anyone—generate such a list, a view about what eudaimonia (as rare as it may be) consists in? The process is one of what Rawls calls "reflective equilibrium." One observes lives that work and lives that don't work to produce what a reflective person thinks is eudaimonia. One examines the conditions of the world and of character that seem to do the job, critically evaluates these with that initial reflective conception of eudaimonia in hand, and adjusts one's conception of both the "conditions" productive of and/or constitutive of eudaimonia and one's conception of 'true happiness/flourishing' along the way. Close inspection begins with an initial hypothesis about what eudaimonia is, what leads to it, what constitutes it, and so on. Call the starting conception eudaimonia$^{\text{initial reflective conception}}$. One deploys eudaimonia$^{\text{initial reflective conception}}$ in examining the roots, sources, and character of eudaimonia$^{\text{initial reflective conception}}$, and this process yields a more sophisticated view; call it eudaimonia$^{\text{mature reflective conception}}$. The "more mature reflective conception" is better, but there is no guarantee it is the best picture we can come up with.

I said that eudaimonics is empirical. Here, first pass, is why: One starts with a hypothesis about what constitutes a healthy or good person (compare to botany on what a healthy plant of a particular variety is), and one asks questions about what causes and constituents contribute to or make up the well-functioning form (how much sun, how much water, what soil type(s), and so on are necessary). One tests these ideas and reaches a well-confirmed hypothesis, possibly adjusting one's initial conception (these plants look very beautiful, but they are short-lived, being susceptible to a certain fungus).

Eudaimonics after Aristotle

In *The Therapy of Desire* (1994), Martha Nussbaum presents a compelling case for understanding the post-Aristotelian Greek and Roman philosophers as doing much more than simply advancing and refining Aristotle's ethics. They advanced eudaimonics by explicitly endorsing the medical analogy. Furthermore, post-Aristotelian ethics advances a view of the good life that is open to everyone, not just the well bred. Despite the universal access to a life of virtue, the Epicureans and the Stoics[2] paint a more demanding picture of virtue than Aristotle does. This more demanding ethical conception requires much deeper psychic change than Aristotle thought necessary in order to alleviate suffering and bring happiness in its place. The need for greater direct attention to an individual's psychic economy is due in part to the fact that Aristotle was insufficiently attentive to the way certain destructive states of mind, for example greed and avarice, cause suffering and bad actions but are nonetheless subject to voluntary control. According to the therapists of desire, more than good socialization, even as supplemented by attending Aristotle's lectures on ethics, is required for virtue. Direct therapy on the minds of adults to quell or eliminate negative desires is needed as well. In addition, the expansion of the list of virtues to include universal compassion requires work to expand and enhance whatever tendencies of fellow feeling are rooted in our nature but are enhanced insufficiently by local (Aristotelian) moral conventions.

Fortunately, the transformation of the psyche required for true virtue and happiness is possible so long as the philosopher equipped with a more expansive set of instruments than argument alone plays the role of a trainer or physician for the soul: "The Hellenistic philosophical schools in Greece and Rome—Epicureans, Skeptics, and Stoics—all conceived philosophy as a way of addressing the most painful problems of human life. They saw the philosopher as a compassionate physician whose arts could heal many pervasive types of human suffering." (Nussbaum 1994, p. 3) These Greek and Roman compassionate philosophers—among them Epicurus, Lucretius, Zeno of Sidion, Chrysippus, Pyrroho, Seneca, Cicero, Epictetus, Sextus Empiricus, and Marcus Aurelius—were founders, luminaries, and practitioners of schools that arose after Aristotle's death (322 B.C.E.) and that remained highly influential into the second or the third century of the Common Era.[3]

The therapists of desire provided (indeed insisted on providing) strong—ideally, valid and sound—arguments to support their diagnoses, prognoses, and therapeutic practices. In part this was because, like their Indian counterparts, especially Buddhists, they believed that mistaken views are often at the root of human suffering—for example, money is widely thought to bring happiness but doesn't.[4] But in contrast to the Hellenistic therapists of desire, they also recognized that argument alone does not always produce the necessary change. Even if false belief—what Buddhists call delusion or wrong view (in Pali, *moha*)—is lifted at some conscious level ("OK, money doesn't bring happiness; I get it now"), there are typically longstanding emotional and wider conative tendencies and attitudes associated with the false belief (possibly antecedent to it) that, in virtue of being deep-seated and partly unconscious, may still control the motivational circuits. Even if the false belief is exposed as false, acquisitive desires and behavior may not abate ("I know that money doesn't bring happiness, but I keep trying to accumulate wealth, and I feel vacant, empty, dissatisfied"). Here the therapists of desire rightly saw the need to bring to bear techniques, in addition to arguments, to adjust or change the economy of desire, often working to eliminate certain destructive emotions by antidotes that were psychologically incompatible with them. Suppose you really get that the arguments or reasons for why you should not have or continue to have sex with a certain object of your lust, but still you can't stop. One antidote is to imagine the object of lust dead and decomposing. Unless you're a necrophiliac, this may work to diminish the grip of the lust. Michel Foucault refers to this style of doing philosophy, which involves working to form or restructure the self, as utilizing *techniques de soi* (techniques of self-work). Nussbaum agrees, but she warns that, then as now, there were *techniques de soi* that relied on mesmeric force and hocus-pocus without the requirement that sound arguments also be offered warranting soul change of a particular sort by way of a suitable technique.[5]

The Moral Psychology of the Buddhist *Abhidhamma*[6]

In moral conformity to the "Four Noble Truths" that guide all Buddhist practices, the Dalai Lama writes: "The principal aim of Buddhist psychology is not to catalog the mind's makeup or even to describe how the mind functions; rather its fundamental concern is to overcome suffering, especially psychological and emotional afflictions, and to clear those afflictions." (2005, pp. 164, 165) This is true, although Buddhist psychology contains a remarkably complex taxonomy and examines the causes and the consistency of the states it taxonomizes. To the puzzlement of many Westerners, the *Abhidhamma* taxonomizes mental states into wholesome and unwholesome (and, to a lesser extent neutral) kinds. Occasionally I mischievously try to get a rise out of scientific friends by telling them that this is so and that I approve of it. This always works. A typical response is "Owen, have you lost all sense of standards?!" Many colleagues think it shockingly irresponsible and very naughty to mix scientific psychology with ethics. In reply, I ask "Have you read any books on abnormal psychology or psychiatry?" The point as I see it is this: If we criticize a psychiatry text on the grounds that it operates with an "unwarranted conception" of mental health, the burden is on the critic to explain why the conception is unwarranted. Likewise, any concerns with the ascriptions of "wholesomeness" or "unwholesomeness" require showing what is wrong in the Buddhist conception of the good life. Is the problem simply doing psychology in a way that is normative? Or is it particular ascriptions of wholesomeness or unwholesomeness that are thought to be unwarranted? The point is that the critic must provide arguments, especially if she judges that there is something deeply wrong in the very enterprise of eudaimonics. The critic must put up or shut up. The fact that the *Abhidhamma* combines descriptive with normative insights gathered from the Buddha's teachings is not itself an objection of any sort so long as the normative claims can be supported by evidence that embracing them captures worthy aims and that abiding by them increases the chances of achieving whatever good it is that the norms point toward.

In any case, the eudaimonistic impulse of Buddhist psychology cannot be emphasized enough.[7] The *Abhidhamma* is a masterpiece of phenomenology. And despite what the Dalai Lama says about not being concerned with taxonomy, the *Abhidhamma* remains arguably the best taxonomy of conscious mental state types ever produced. Most of my remarks pertain to books 1 and 7, which deal with psychology. Buddhists treat books 1 and 7 of the *Abhidhamma* as a psychological masterpiece combining deep phenomenology, analytic acuity, and classification of mental states in terms of the "wholesome" and the "unwholesome" in accordance with how they fit into the Buddhist view of eudaimonia.[8]

What first catches the eye of the Western reader is the extraordinary number of distinctions drawn among states of consciousness. The book begins with a taxonomy of consciousness (*Citta*) into types of conscious mental states (*cittas*). These number 89 initially, and after some adjustments they reach 121. Each type is characterized in terms of the sort of object it takes in (so visual and auditory consciousness differ in an obvious way); its phenomenal feel (e.g., sad or happy), its proximate cause or root (e.g., there is greed-rooted consciousness and hatred-rooted consciousness), and its function or purpose (avaricious consciousness is thirsty and aims to suck in and swallow what it desires).[9]

As one studies the *Abhidhamma*, one gets into the spirit of drawing distinctions upon distinctions. Indeed, one could really start to believe the Tibetan joke that a master phenomenologist might be able to discern 84,000 (the number is akin to "a gazillion") types of anger or craving. The second thing (or perhaps it will happen first) that will strike the Western reader is that the words 'wholesome' and 'unwholesome' (and, less frequently, 'neutral') are used in the process of classification itself. Again one can easily imagine the objection that "real" (i.e., "truly scientific") psychology describes, explains, and predicts, but does not judge, the various kinds of sensations, perception, emotions, and learning it analyses in normative terms. This is a correct description for the current state of play as regards how disciplines are to behave, but it says nothing principled about why it should be understood as an objection about the epistemic legitimacy of engaging in normative discourse and reaching normative conclusions.

In any case, the three poisons that come with our natures are hatred (*tanha*), craving (*lobha*), and delusion (*moha*). These three, uniquely perhaps, are always bad or unwholesome. Furthermore, they ramify and interact with other mental states, indeed with one's overall sense of well-being, in ways that produce unease.

The three poisons are first elaborated as giving rise to "The Six Main Mental Afflictions": Attachment of craving, Anger (including hostility and hatred), Pridefulness, Ignorance and delusion, Afflictive doubt, and Afflictive views. These are roots of the "Twenty Derivative Mental Afflictions." Anger comes in five types: Wrath, Resentment, Spite, Envy/Jealousy, and Cruelty. Attachment also comes in five types: Avarice, Inflated self-esteem, Excitation, Concealment of one's own vices, and Dullness. There are four kinds of Ignorance: Blind faith, Spiritual sloth, Forgetfulness, and Lack of introspective attentiveness. Finally, there are six types of Ignorance + Attachment: Pretension, Deception, Shamelessness, Inconsideration of others, Unconscientiousness, and Distraction. The point is that there are a lot of ways one can go wrong. The tools required for the therapy in virtue will, not surprisingly, have to be abundant and multifarious.

At the other end of the spectrum from the poisons and the derivative afflictions are the "Four Divine Abodes" (*brahmaviharas*), also called the "illimitables" or "immeasurables" (*appamanna*): Loving-kindness (*metta*), Compassion (*karuna*), Appreciative Joy (*mudita*), and Equanimity (*upekkha*). Any person who takes the bodhisattva's vows in the privacy of his or her mind—and anyone can do so without ceremony, pomp and circumstance —is on the way to embodying these virtues. The vow is to work to liberate all sentient beings from suffering (the work of "compassion") and to bring happiness in its stead (the work of "loving-kindness").

One might think that, just as the "three poisons" are categorically bad, the "divine abodes" are categorically good. Maybe. But an important caveat is in order. Wisdom (*panna* in Pali; *prajna* in Sanskrit) and virtue (*sila*)— require as a necessary condition avoiding the three poisons. It seems constitutive of loving-kindness, compassion, appreciative joy at the success of others, and equanimity that these states rule out (are psychologically incompatible with) hatred and envy (a form of greed). But delusion (*moha*) can mitigate the sublimity of even the "divine abodes." Suppose one achieves equanimity because he fails to notice certain particulars about his own character, or the character or plight of others, that he ought to be noticing or paying attention to. Here delusion surfaces and might make us question whether the equanimity is warranted. It feels sublime, but it is supported and sustained by failing to see what one ought to see. There is an unwholesome aspect to such equanimity.[10]

For reasons such as these, Buddhist psychology pays a considerable amount of attention to the *causes* of mental states, especially before moral assessment is made, as well as to the causes of epistemic disruption (internal and external) and to matters of intention and moral responsibility. All states rooted in hatred or greed are unwholesome, as are states caused by wrong views for which I am responsible. The case of the magic pill that makes one happy is not discussed, but similar cases are. If a seizure causes me to experience euphoria, it is deemed "rootless," and rootless states are unwholesome.[11] Notice, this means that one can be in a state that feels good, has positive valence, but is nonetheless unwholesome. One reason for judging the state unwholesome is that it is normative—both psychologically and ethically—that happiness be produced by goodness I possess or self-work I engage in, not by aberrant or undisciplined neural firings over which I have no control.

All four abodes are said to involve states of mind toward others. One might agree while emphasizing that, at the same time, all four are in fact states of mind of the individual who has them, and they have unique first-person phenomenological feel for that person. Their object is, of course, the good of some other. But this analysis seems to run into trouble with equanimity (*upekkha*), which might seem to be a pure state of my soul, and thus not directed at, for, or toward anything outside me. To be sure, my being calm and serene might make me more pleasant to be around or more caring toward others, but it is not constitutive of equanimity, as English speakers understand 'equanimity', that it has this aim or quality. It is sufficient for equanimity that I am serene in and toward myself, as it were.

This is not how Buddhists understand equanimity. Equanimity (*upekkha*) means more than serenity. It is constitutive of equanimity that I feel impartially about the well-being of others. If I am in the state of equanimity, interpreted as *upekkha*, I am in a state that involves, as an essential component, equal care and concern for all sentient beings.[12] We might translate *upekkha* as "equanimity in community" if it helps us avoid confusion with our understanding of equanimity as a purely self-regarding state of mind.

The "Internalist Objection" to Eudaimonics

Consider an objection that one might make (I would) about the two eudaimonistic views I have sketched. The objection is this: Suppose one accepts the idea that the work discussed has empirical grounding. There is in Aristotle observation of how the "many" and the "few" are, and how they fare, governed by the method of "reflective equilibrium." Buddhism, I claim, operates with the same sort of methodology. Once the Aristotelian and the Buddhist articulate their conceptions of an excellent life, no answer is given as to which of the two contenders is the right one. Nor can such an answer be given, the critic claims. Why? Because the only measure of what constitutes eudaimonia, flourishing, excellence, enlightenment is in fact what norms are avowed, practiced, considered best from *inside* the culture. Who had it more right: Newton, or Einstein? Easy. Science resolves such questions. Who gives a better picture of eudaimonistic excellence: Aristotle, or Buddha? Normative mind science can give no answer. The Aristotelian answer is right for Aristotelians, the Buddhist view for Buddhists. What can the eudaimonistic empiricist do?

Let me make the worry worse before making it better. Consider two relatively recent statements by the Dalai Lama:

> Now, we are made to seek happiness. And it is clear that feelings of love, affection, closeness, and compassion bring happiness. (1998, p. 52)

> For our life to be of value, I think we must develop basic good human qualities—warmth, kindness, compassion. Then our life becomes meaningful and more peaceful—happier. (ibid., p. 64)

There is a reading of both these statements as empirical, that is, as statements about what sort of things produce eudaimonia.[13] Reading these statements as "empirical" is to be contrasted with reading them as mere endorsements or recommendations for living a certain way. (I think they are both.)

Now, if the empirical claims[14] were true, a sensible person would want to adopt certain norms. The reason is simple and straightforward: Adopting the relevant norms—for example, living in a loving, compassionate way—

will bring happiness and meaning (on the assumption that they are claims that living in certain ways is necessary for producing happiness and/or meaning). The idea is to read the Dalai Lama's two statements as truths of a universal philosophical psychology. It is true at all times and places that love, compassion, and their suite bring happiness and meaning, assuming that other things are in place (that one has enough to survive, that one is not constitutionally insane, and so on).

But consider this possibility: Once any culture, social group, or moral or spiritual tradition has settled on a certain conception of eudaimonia$^{\text{mature reflective conception}}$, "the tradition" is almost guaranteed to create a social environment that pulls for endorsement of the virtues, values, and norms said to enable, cause, or constitute eudaimonia$^{\text{mature reflective conception}}$. This will make it true that to find genuine happiness you need to be a certain way and do thus and so—but not because being and doing thus and so produce "true happiness" for everyone; rather, because the contingencies of social reinforcement around here are such that only in this way are happiness and meaning granted. This is what I call "the internalist objection."[15]

At this point. as worries about the imprecision and circularity begin to intrude, one usually quotes a certain passage from Aristotle, perhaps to divert attention. I dutifully do so, by way of my own free translation:

> We should consider our discussion adequate insofar as we make things perspicuous enough as regards our subject matter. We do not seek or expect the same degree of exactness in all sorts of arguments (compare: mathematics, physics, history), just as we do not expect exact sameness in the products of different crafts (compare pressing coins, to knitting clothes).... In ethics and political science each of our generalizations ought to be understood as holding true usually. And because this is the nature of our premises [that such and such holds generally, but not universally], we must be satisfied with probabilistic conclusions of the same sort. (*Nicomachean Ethics*, I, 2, 1094b–1095a)

Aristotle's aim is to distinguish the inferences of mathematics and physics from eudaimonistic scientia (which we now see includes both ethics and political science). Mathematics involves deductive arguments and yields necessary conclusions. Physics starts with universal generalizations: Always and everywhere and for all things, $f = ma$.[16] Arguments with the latter as a major premise can yield a certain conclusion if and only if two of the three variables can be specified exactly. If not, we deal with approximations, and our conclusion yields the same.

The first point about eudaimonics is that it has no universal generalizations. The reason(s) could be practical, epistemic, or ontological. Aristotle is a naturalist so I think it best to understand him as thinking that the lack of universal generalizations is a practical and/or epistemic matter. It would be false to say "Money never brings subjective happiness." Why? Because given our practical methods for divining states of subjective happiness, money does appear to make some people happy.[17] One might make this move: Even if money sometimes brings a high and sustained state of *subjective happiness* it never brings *true happiness*. Here, though, one might legitimately worry that we have imported a stipulation to the effect that if happiness is caused by money, then it isn't to be counted as true happiness. And this would be a mere exclamation of endorsement for a particular way of speaking and moralizing. It would not, as I imagined the situation, involve making an empirical claim. That said, if the proponent of the stipulation predicted that when neuroscience matures and we can more precisely state what neural state is a state of subjective happiness and which is a state of true subjective happiness, then those few who money seems to make happy will not show up in the class of those whom we deem *truly* happy. Then we would be able to state that "Money never brings true subjective happiness." At present, the imagined scenario is pure science fiction.

It is a fact that most of the Forbes 500 richest Americans say that they are happy. But the question of what causes their happiness is well nigh intractable. First, these individuals can't do a phenomenological analysis of themselves and say what and to what degree various factors contribute to their happy state. Third-person techniques are also not up to the task of distinguishing among all the causal contributors to these individuals' states of mind. So although money does for some look as if it is the main cause of their subjective happiness, it may in fact be caused by genes, unusual maternal care, or by the fact that they always get enough sleep, or by a whole bunch of such things (including the money).

Whatever the cause, we should just accept that most "laws" in *Geisteswissenschaften* are statistical generalizations, not universal laws. This doesn't make any findings less empirical, it just makes them less precise.

With this much in hand, let's return to the "internalist objection." The objection can now be stated this way: OK, we'll grant you imprecision: your science is immature[18] and your domain of inquiry is very complicated. What we won't give you is this: you can't give any external, objective reasons for your judgments about what states of affairs, norms, and ways of life are best or most worthy that doesn't privilege a class of norms or ways of life that are part of your data set. You mix norms with facts.

This is true, at least to a point. The question is, should it worry us? Note we do not commit Hume's fallacy. There is nothing in either Aristotle's approach or the Buddhist approach that involves attempting to *derive* oughts from is's, or values from facts. This is because deductive logic is not in play. In fact, it is my opinion that in the history of ethics (almost) no one has been stupid enough to actually commit Hume's fallacy, although some people smuggle oughts in a bit too easily without adequate argument.[19]

The reasoning involved is inductive, abductive, statistical, and probabilistic. And most of science except (possibly) theoretical physics is like this. Recall the method of "reflective equilibrium" as described above. Rawls, as I said, names his method as such, but it was Aristotle who first described how it works: We survey the views and aspirations of the "many" and the "wise" (1098b27, 1153b31, 1173a14, 1179a16). We treat all these beliefs and ways of life as worth taking seriously, but as revisable (1095a22, b16 1113a33, 1153b35, 1159a19, 1163b26, 1167b27, 1172b3). We use techniques such as these to uncover problems with the beliefs of both the many and the few: do people who think money brings happiness, usually gain it? No. Instead they get stuck on what we nowadays call the hedonic treadmill. Is honor satisfying? Yes, but normally only if deserved, and so on.

We begin responding to "the internalist objection" by admitting we are looking for norms, values, and practices that are the best, where "the best" is almost always "the best for such and such purpose or purposes." The analogies are from engineering and the crafts. Given that we need/want bridges that work and shoes that don't leak, we cull from all the possible ways of accomplishing these things the ones that work best. Were Athenian bridge builders or Athenian cobblers the best ever? Of course not. But they were the best at that time.

The situation is Kuhnian (1962, 1970). Just as a scientific theory stands as best until there is a competitor that does better, so with bridge-building, shoe-making, and ethics.[20] At this point the comparative ethicist can usefully chime in. One worry I gave voice to (although, truth be told, it doesn't normally worry me much) is this: Aristotle offers one set of virtues and the Buddha another. Perhaps they are not inconsistent, but they are not the same. And so far no answer has been given as to which is best. The critic of eudaimonistic scientia will say that this problem is to be expected, indeed it is inevitable, given the objectionable internalist methodology being deployed.

Here is a sketch of a reply: First, we are not required to simply accept Aristotle's picture because he offers it as his best theory—even as the correct theory for Athenians. Take his last two virtues, magnanimity and great-souledness. The latter might be criticized, or at least warrant watchfulness, because it seems as if the great-souled person might more than others be prone to self-puffery. Aristotle is careful to say that the expectation of honor is to be based on correct perception of merit. But given that *hubris* is bad, we are entitled on grounds internal to the Greek situation to worry about the virtue's stability. Regarding magnificence, we have resources for criticism that are both *internal* and *external*. What I mean is this: In order to be magnificent in Aristotle's sense, requires that one be rich. One might think wealth comes from merit. But given the attention paid by the Greeks to luck in life's circumstances, one might see that wealth is often a matter of luck (*eutuches*). One response is to say that this is why magnificence is a virtue, the excellent person gives away a lot of his stuff for the common good, both because this is good, and because it involves recognition that his wealth is not a measure of his merit—it comes from a lot of good luck. These critical resources are internal to (but perhaps not fully visible in) the tradition. But again suppose that the matter of luck is just not connected in the minds of the Athenians to wealth distribution. The internal dialectic doesn't, for whatever reason, let those who regularly think about "luck" see its role in the case of wealth.

Here we can go external: First, there is the small matter of the economy built on slaves (actually, at this time, usually prisoners of war). Aristotle is silent on the legitimacy of slave ownership; it is part of the-taken-for-granted background. So we raise some questions from outside about that practice. The most promising external tactic, however, would be to point to economic arguments about how wealth accrues. Wealth acquisition does not reliably track merit, and it can, if we are not watchful, lead to sufficiently uneven distributions such that there are prudential reasons (safety and security given that the poor and oppressed might rise up) to do some evening out. If we think there is also ethical knowledge, we can form a tandem. Think of Amartya Sen's work: we now know enough about ethics and economics to know that the distribution of wealth can and should be reconfigured if the wealth of the few is based on mistreatment or oppression of the many. Aristotle didn't see or know this. We do. And thus when his list of virtues is placed into "wider reflective equilibrium," we can make a normative recommendation that does not challenge perhaps the virtue of magnificence, but that does challenge the social conditions that make the virtue one that only a few can display.

The strategy of seeking "wider reflective equilibrium" can now be used in comparing Aristotle's conception of eudaimonia with the Buddhist conception. I simply won't discuss the possibility that the views might be incomparable or incommensurable. I don't think they are, although they definitely have features that make most sense internal to the traditions themselves. And this is predicted by conceiving of ethics as a form of human ecology, a line I have been pushing for over a decade. The idea is that there are almost certainly universal necessities across all human environments that pull for and thus make rational certain prohibitions such as ones against murdering innocents, stealing rightfully owned property, and so on.[21] Beyond these "big-ticket" items, local ecological conditions will create their own pressures on normative construction (Flanagan 2002; Wong 2006).

In the case of the Aristotelian-Buddhism comparison we can note first that being asked to compare them puts us in Kuhnian space. Neither theory, taken on its own, we will suppose, had in its time and place a serious, worthy competitor. So we put them into competition across space and time. Here—without fully engaging in the exercise—is what we should expect to find in such a competition. We will find that both theories have strengths and weaknesses that one only sees—or sees best—when they are discussed together. I can't and thus won't elaborate here, except to say this much. If we take everything on Aristotle's list, we see most of his virtues in the *Abhidhamma* as wholesome states. What we see in Buddhism, but not in Aristotle, are the four divine abodes. If you think as I do that Jesus of Nazareth and the Buddhist tradition (possibly Mencius, and even Mozi, in China) saw the merit of expansive love and compassion in ways that Aristotle didn't (because of how he was positioned internally to the Athenian ecology of value), then we can recommend expanding Aristotle's list. Examined from a perspective of wider reflective equilibrium, the Buddhist virtues can be seen as excellent and thus as worthy of being added to Aristotle's list.[22]

This, then, is how I respond to "the internalist objection." I concede that we are engaged in an empirical inquiry that involves methods not normal or required in what we think of as observer independent sciences where normative questions don't arise (as visibly). All the human sciences are reflexive. We do the inquiring and we are the objects of inquiry. This is just a fact about what has to happen if you want to study humans. One important task is to describe and explain what goes on in perception, emotion, and cognition. Another task is to answer overtly normative questions about the nature, causes, and constituents of eudaimonia. Such normative inquiry has very precise analogies in sciences such as engineering and botany. Given certain ends, how best can we achieve them? Engineering exists by combining knowledge from physics, material science, and so on, with the intention to satisfy non-frivolous practical desires. Botanical and agriculture science work the same way (botany, of course, incorporates aesthetic values). Same with eudaimonics. The aims to make sense of things and to live meaningfully are perfectly natural and legitimate. How best, given our situation, can we best satisfy these ends? In all the latter cases, the method of reflective equilibrium is the only method ever proposed to do the job.

One might think that the use of the method cross-culturally is a way of testing in ethics what in science would be called "rival theories." And it is. But there is still a difference that makes a difference. Where there are sociomoral or political problems or situations that afflict life in both cultures, and where one theory does better than the other at solving it, then and only then is the situation one in which both sides can/might agree that the rival is more resourceful (MacIntyre 1981, 1988, 1990). In science the cases of "crucial experiments" are almost always ones in which two rival theories predict inconsistent objective outcomes (light will curve or it won't when the eclipse occurs). One might claim that ethics doesn't engage in prediction to the same degree that normal science does. Perhaps. But we do, I claim, underestimate the role that predictions about eudaimonia play in ethics.

Still, the critic might complain that in ethics there are two familiar sorts of cases that don't seem to arise in normal science and which militate some against the idea that there could be anything like a crucial moral experiment that could decide between two rivals. First, sometimes "the problem" that might reveal the strength of one theory simply isn't a problem from another culture's perspective. Genetic engineering is not an issue on the radar of most people in the Himalayas. They need cataract surgery, not gene tampering. Another familiar kind of case is one in which deep-seated and non-shared background assumptions logically necessitate different answers to the same moral question.

Imagine that we ask former communist Chinese thinkers and Western libertarians "How unfettered ought capitalism to be?" or that we ask Dutch secularists and Roman Catholics "Is active euthanasia morally acceptable now that we have very merciful techniques?" In both cases, we know that unshared background norms will profoundly color moral judgments about how to deal with the technology, economic practice, or whatever. In such cases we can rightly locate the source of disagreement in the background norms that are taken-for-granted. But there will not be world enough and time to revisit and examine these. Furthermore, if one is dealing with a problem for which a particular ethical tradition has its own precious answer the need to probe that deeply simply won't arise.

Here again we seem to be hitting upon a methodological difficulty that could make ethics look "methodologically infirm" (Quine's view) compared to what we normally call "science." Normal sciences we might think yield universal truths. Could moral knowledge—unlike, say, knowledge in physics—be local knowledge? The answer depends on what moral knowledge consists in and what it is for, what purposes it serves.

Naturalism, Ecologies of Virtue, and Comprehensive Pictures of "The Good"

In the two passages quoted here, John Dewey captures perfectly what my kind of naturalism in ethics commits us to:

> For practical purposes morals means customs, folkways, established collective habits. This is a commonplace of the anthropologist, though the moral theorist generally suffers from an illusion that his own place and day is, or ought to be, an exception. (1922, p. 55)

> Moral science is not something with a separate province. It is physical, biological, and historical knowledge placed in a humane context where it will illuminate and guide the activities of men. (ibid., pp. 204, 205)

Morals consist of habits of heart, mind, and behavior. Morality is "normative" in the following sense: It consists of the extraction of "good" or "excellent" practices from common practices. Ethics consists of wisdom based on historical experience about how best to arrange our affairs, and how to develop our nobler potential, as this too is judged on the basis of historical experience. Moral habits, wisdom, and skills consist largely of "know-how" that allows for smooth interpersonal relations, as well as for personal growth and fulfillment. Ethical reasoning is a variety of practical reasoning designed to help us negotiate practical life, both intrapersonal and interpersonal, as it occurs in the ecological niche we occupy. We do not know *a priori* how to be, how to live. We are social animals. One thing this means is that we must be immersed in a culture and a Space of Meaning if we are to make sense of things and find meaning.

Moral knowledge is not a kind of "divine wisdom." The naturalist has no truck with those who claim that some moral norms have a supernatural warrant, or that the reason to be good is for the sake of reaping a heavenly reward. We are animals. This world is a material one, and there is no justification, none whatsoever, for believing in divinities or an afterlife. Morals matter, but they can't really be about or for the sake of what the non-naturalist says they are about or for. Morals are not about what God thinks is good or even what God commands, nor are morals about serving God's purposes or doing God's will. These ideas are out there, but they are childish ideas that are epistemically unwarranted.

If ethics is about anything it must be about flourishing. Supposing this is true, what if anything contentful can be said about human flourishing—about eudaimonia? What progress can be made in securing either or both of the following theses (or, what is different, making them credible)?

(1) It is a necessary condition of *subjective flourishing* that the virtues an individual displays and norms she avows and abides pass tests for reflective equilibrium;

(2) It is a necessary condition of *objective flourishing* that the virtues an individual displays, and the norms she avows and abides pass test for wide reflective equilibrium.

Reflective equilibrium (RE) differs from wide reflective equilibrium (WRE) in this way: A moral conception passes tests of reflective equilibrium if one lives well and is good as these are conceived by one's culture. A moral conception passes tests of WRE if the conception passes tests of inter-cultural comparison, in particular if it passes tests that seriously consider credible alternative conceptions of the good. RE is an internal way of testing a moral conception; WRE is external in one familiar sense.

There is something close to a consensus among psychologists who work on moral psychology that some form of social intuitionism is true and that intuitions are constructed internal to a normative conception. Intuition is defined this way:

> Intuitions are the judgments, solutions, and ideas that pop into consciousness without our being aware of the mental processes that led to them. When you suddenly know the answer to a problem you've been mulling, or when you know that you like someone but can't tell why, your knowledge is intuitive. Moral intuitions are a subclass of intuitions, in which feelings of approval or disapproval pop into awareness as we see or hear about something someone did, or as we consider choices for ourselves. (Haidt and Joseph 2005)

Many will say that evidence from brain imaging shows that intuitions are limbic responses, not higher cortical ones (Greene et al. 2001; Greene and Haid 2002; Greene 2006). The reason is that morality is built on top of certain initial emotional settings. This fact about how mature agents check and/or resolve moral problems on-line does not mean that the original rationale for a moral response is emotive. I don't go into that matter directly here, but my discussion of RE and WRE assume that there is a distinction between on-line moral response and considered judgments about the worth of virtues, norms, and values. Even if the moral community secures rapid moral judgment by utilizing quick emotional response in mature agents, it most likely trained novices by establishing linkages between cognition and action: "This is a situation of such and such a kind, and it deserves this sort of rapid response."

In the ethnographic and psychological literature, four domains of life appear universally as socio-moral domains that human build normative space around. The moral community works at coordination between four classes of emotions (compassion related, resentment related, anger related, and disgust/contempt related) *and* four domains of life (suffering, power, exchange, and purity). The coordination between the four families of emotions and the domains were enabled, we assume, originally by evolution. These domains are suffering/compassion, reciprocity/fairness, hierarchy/respect, and purity/impurity.[23]

Table 4.1

Four moral modules and the emotions and virtues associated with them. Source: Haidt and Joseph 2004.

	Suffering	Hierarchy	Reciprocity	Purity
Proper domain (original triggers)	Suffering and vulnerability of one's children	Physical size and strength, domination and protection	Cheating vs. cooperation in joint ventures, food sharing	People with diseases or parasites; waste products
Actual domain (modern examples)	Baby seals, cartoon characters	Bosses, gods	Marital fidelity, broken vending machines	Taboo ideas (communism, racism)
Characteristic emotions	Compassion	Resentment vs. respect, awe	Anger, guilt vs. gratitude	Disgust
Relevant virtues	Kindness, compassion	Obedience, deference, loyalty	Fairness, justice, trustworthiness	Cleanliness, purity, chastity

The well-confirmed working assumption is that the social life of *Homo sapiens* and of our ancestors requires attention to these domains because issues of power, suffering, family relations, and hygiene require attention. It is essential to understand that these tendencies, conceived as "reactive attitudes" (Strawson 1965; Flanagan 2002, 2003b) originated in distant ancestors, not *de novo* in the lineage of *Homo sapiens*. Indeed, Darwin (1871, 1872) saw both facial and behavioral homologues of some of these attitudes in canines. Presumably the dispositional modules with their associated emotions evolved as adaptations in ancient environments to solve common (and possibly universal) adaptive challenges.

The modules are universal. But that is the level at which universality in a strict sense ends. The reason is this: different natural and social ecologies pull for different solution strategies to local adaptive challenges. Differences among social landscapes and ecologies, differences among starting points, helps explain the plurality of moral conceptions (Wong 1984, 2006).

Most relevant for eudaimonics, it would be a very big mistake to think all normative solutions to adaptive problems are designed with the end of widespread human flourishing in mind, let alone with the flourishing of all sentient beings in mind. If I am well off, I may endorse norms to keep my self and my loved ones that way and keep those worse off in their place. If I have the power to endorse those norms and make the common folk think they are divinely inspired, I and my people are "winners" and the "losers" think things are as they are supposed to be. Thrasymachus in Plato's *Republic*, Marx, and Foucault think this mode of normative design is quite common. It is something to be watchful for.

The overall picture from social intuitionism is this: we have evolved with a set of dispositions to respond intuitively by moralizing these four (or more[24]) domains. Haidt thinks some of the moralities that develop on top of the four dispositional modules are "incommensurable." I am not sure. Even something less that "incommensurability" would help explain moral conflict. And either way there will be problems of moral communication.

One might, adopting language from John Rawls, think of our expansive moral conceptions as involving "comprehensive conceptions of the good." Ethics, politics, and spiritual and aesthetic traditions interact to provide the basic structure of a comprehensive conception, and this basic structure determines for most people how they conceive of life within the Space of Meaning$^{\text{Early 21st century}}$. Aesthetic attitudes, and attitudes toward science and technology, will be visibly affected by how persons locate themselves within the spaces they see as providing sense and meaning. And it is worth emphasizing that there will ordinarily be some division of labor to resolve conflicts among norms that have their primary homes in spaces of meaning that pertain to ethics, often conceived as personal, and political and economic norms typically understood to pertain to life conceived more widely, involving multi-person or group coordination and cooperation patterns.

Among Americans, there is evidence for differences in moral habituation between politically liberal and politically conservative adults, which explains differences in moral attitudes over generations. One way in which cultures vary is by differentially activating, growing, or tamping down the initial settings of the four modules:

> American Muslims and American political conservatives value virtues of kindness, respect for authority, fairness, and spiritual purity. American liberals, however, rely more heavily on virtues rooted in the suffering module (liberals have a much keener ability to detect victimization) and the reciprocity module (virtues of equality, rights, and fairness). For liberals, the conservative virtues of hierarchy and order seem too closely related to oppression and the conservative virtues of purity seem to have too often been used to exclude or morally taint whole groups (e.g., blacks, homosexuals, sexually active women). (Haidt and Joseph 2005)[25]

The larger positive-psychology movement typically identifies important virtues and then lists "strengths of character"—ways the virtue is realized, or (perhaps better) tools differentially required for displaying or embodying the virtue. Recall from chapter 2 that the positive-psychology movement (one major strand, at any rate) claims that these virtues are universal[26]: Wisdom, Courage, Humanity, Justice, and Temperance. For each major virtue, there is a set of "strengths of character," which typically enable or embody the virtue. The "strengths of character" for the virtues Humanity, Justice, and Temperance are the following (Peterson and Seligman 2004, pp. 29–30):

Humanity

Love: valuing close relations with others, in particular those in which sharing and caring are reciprocated; being close to people

Kindness [generosity, nurturance, care, compassion, altruistic love, "niceness"]: doing favors and good deeds for others; helping them; taking care of them

Social intelligence [emotional intelligence, personal intelligence]: being aware of the motives and feelings of other people and oneself; knowing what to do to fit into different social situations; knowing what makes other people tick

Justice

Citizenship [social responsibility, loyalty, teamwork]: working well as a member of a group or team; being loyal to the group; doing one's share

Fairness: treating all people the same according to notions of fairness and justice; not letting personal feelings bias decisions about others; giving everyone a fair chance

Leadership: encouraging a group of which one is a member to get things done and at the same time maintain good relations within the group; organizing group activities and seeing that they happen

Temperance

Forgiveness and mercy: forgiving those who have done wrong; accepting the shortcomings of others; giving people a second chance; not being vengeful

Humility/Modesty: letting one's accomplishments speak for themselves; not seeking the spotlight; not regarding oneself as more special than one is

Prudence: being careful about one's choices; not taking undue risks; not saying or doing things that might later be regretted

Self-regulation [self-control]: regulating what one feels and does; being disciplined; controlling one's appetites and emotions

I don't know if Peterson and Seligman would approve of this interpretation, but here is one way to think of their lists of "virtues" and "strengths of character": "Virtues" name virtues at a superordinate level; "strengths of character" name "virtues" at a lower level. Courage is a superordinate virtue; in one place and time it will be embodied primarily as a military virtue, in another place and time as a form of "integrity" (say, standing up for one's convictions). With regard to the three virtues picked out above—Humanity, Justice, Temperance—the strengths of character read my way explain such things as the following:

- Humanity can be wide or narrow, local or global, in-group or out-group. The virtues of empathy, sympathy, fellow-feeling, and compassion are not specified as the virtues they are until a culture gives them form, scope, and range. For example, in Confucianism *ren* (benevolence) is rooted in the family and designed to spread outward. In Buddhism *karuna* and *metta* (compassion and loving-kindness) start as universal values but focus on those close-to-us for familiar practical reasons. The goal, a good loving world, is perhaps the shared end.

- Justice is compatible with hierarchy or egalitarianism. Every kind of modern political culture has a court system that is supposed to operate fairly. Justice in the courts is possible both in democracies and in various non-democratic polities.

- Temperance will depend on how in particular times and place "disgust" reactions are conceived of—whether they have narrow or wide scope, or really simply what their scope is. Polygamy is disgusting in some places, not in others.[27] The use of intoxicants is OK or not, and if OK it is so only to some sensible point.

- The instantiation of specific virtues under the master virtues will interact and mutually inform each other. If the conception of justice is hierarchical, then it may be particularly contemptuous not to show the elders proper respect (China). If humanness is conceived (under a particular conception of justice) to have a very wide scope, involving love for all sentient beings, then one might feel poorly (disgusted or contemptuous) toward family or nation-state chauvinists (Buddhists), unless you are also taught to always override feelings of disgust or contempt with feelings of compassion and love (Buddhism). It is dialectically very complicated.[28]

I started this section with the question of whether there was anything that sensibly can be said about flourishing. I put forward two ideas that were inspired by Aristotle:

(1) It is a necessary condition of subjective flourishing that the virtues an individual displays, and the norms she avows and abides, pass tests for reflective equilibrium.

(2) It is a necessary condition of objective flourishing that the virtues an individual displays, and the norms she avows and abides, pass tests for wide reflective equilibrium.[29]

We see that inside a modern culture—the United States, for example—that there is more that one conception of the good and of what flourishing consists in. Perhaps there is agreement of the list of virtues, but there is disagreement about their scope, when they are warranted, their form, and so on. In cases where there is diversity inside one's culture, going wide can begin at home. WRE involves taking seriously the competitor(s). If one does so honestly one tries to see what the alternative conception sees and why it sees things as it does.

Moral Networks and Moral Progress

Going wide might result in seeing that there are reasons to adjust one's conception, seeing that it makes sense to take account of what one doesn't take account of, perhaps at least not sufficiently. I can bring out this idea by revisiting a debate I had several years ago with Paul Churchland concerning "moral network theory" and the topic of moral progress (Flanagan 1996b; Churchland 1996). Progress is a normative concept and it is related to the kind of evaluation involved in the two claims that I have been trying to show make sense:

(1) It is a necessary condition of subjective flourishing that the virtues an individual displays, and the norms she avows and abides, pass tests for reflective equilibrium.

(2) It is a necessary condition of objective flourishing that the virtues an individual displays, and the norms she avows and abides, pass tests for wide reflective equilibrium.

One might say there are two conceptions of moral progress tied to (1) and (2), respectively. To make things interesting, imagine this: Churchland believes in progress of type 1: Flourishing and moral improvement have only internal criteria. A morally good person works to reach an equilibrium with the norms avowed inside her culture. To the extent that she does so, she is good by the lights of that culture and has reason to feel self-esteem.

It is true on (1) and (2) that a person who is in normative conformity with the values, virtues, and moral practices advocated in his culture will be deemed a good person by that culture and will in all likelihood self-conceive this way. He makes progress at moral improvement so long as he adjusts his values etc. as required internally by his culture. But the reason to advance (2) in addition to (1) is that satisfying (1) is not sufficient for making an all-things-considered judgment that an individual is genuinely good, virtuous, and so on. Nor is adjustment to the logic of normativity inside a culture a reliable mark of moral improvement (read progress). These are strong claims. Let me try to defend them.

Churchland (1989) uses San Diego-style connectionist mental modeling to provide a neural network model of how a "moral system" such as a person acquires moral knowledge. Churchland was, I believe, the first person to apply neural-network theory to explaining how we acquire moral knowledge. Like Aristotle and Dewey, he argues that our moral capacities are instantiated as "skills"—cognitive, affective, and behavioral knowledge and skills—by a complexly configured matrix of synaptic connections. Connectionism models the acquisition of knowledge in terms of dynamic structures of connections and relationships among discrete neural units. Moral learning is the adjustment of the weighted connections that then constitute the acquisition of a new function, involving in the human case, stimuli-that-are-interpreted, feeling/emotions, and action. Some of the patterns activated to secure reliable function represent categories (e.g., "morally significant," "morally insignificant," "morally bad," "morally good"); others involve the acquisition of domain specific skills (what honesty is, how and when honesty is called for, and so on).

Morality is a natural phenomenon to be studied naturalistically. How could it be otherwise? We are animals. Moral habits in virtue of being complex social constructions on top of our basic psychobiological nature are not guaranteed to be "moral" in the sense of being "good." If humans are correctly modeled by connectionist systems, we will learn what we are encouraged to learn and we will admire and avow norms that are socially endorsed. This will normally be enough to generate, secure, and stabilize self-esteem. And doing that will be sufficient to satisfy (1), our norms will be in reflective equilibrium with what our culture deems good and thus will pass the test of reflective equilibrium. The person who embodies the virtues and conforms to the socially supported rules will feel self-esteem and have good social reason(s) to do so. In that sense she flourishes. But consider that many non-Jewish middle-class women in Nazi Germany satisfied (1) by continuing to abide, even "improving" their attention to, domains of life that German women were supposed to attend to. They were excellent wives, mothers, and so on, while their husbands were off fighting in the Second World War. (See Koonz 1987.) Such women had to know about, and in some cases agree with, the views of many of their fellow German Catholics and Protestants, even if we suppose they knew nothing about the actual Holocaust. Did such women flourish? Were they eudaimon? According to (1), possibly. According to (2), no.

A connectionist system learns. Indeed, such a system specializes in learning. How it does so depends on its initial settings, on experience, and on feedback about how well it is catching on to—in this case—socially endorsed moral norms that the system is not strongly disposed to resist acquiring. The initial settings of a human learning device, conceived along connectionist lines, should have certain reactive attitudes set if they are to model us. Thus the system should model as realistically as possible the activation of basic emotions (disgust, fear, anger, and so on) to their typical environmental triggers. Let us call these the *initial settings* (though this is tricky, insofar as some biological settings are developmental—e.g., ones associated with puberty).

How far can we learn to deviate from the initial settings? There is no general answer, since deviation from each reactive attitude setting depends on a host of factors, including the strength of the particular initial settings. It just may be that a plausible set of initial settings as inspired by evolutionary psychology will make it highly improbable that human females can acquire norms like those of black widow spiders or praying mantises and learn to decapitate their mates after impregnation (forget about the fact that if the guys catch on they will resist). Even if there would be no adaptive loss in terms of sex and child-rearing, the practice might just be too disgusting to catch on.

Churchland has no strong opinion on what the initial settings should be for learning morality—he certainly does not subscribe to a "moral tabula rasa" picture, so I think it would be a friendly suggestion that the initial settings be the ones the social intuitionist proposes, priming in the four intuitive modules for suffering/compassion, reciprocity/fairness, hierarchy/respect, and purity/impurity, as well as (perhaps) in the recently proposed in-group/out-group module.

My initial disagreement with Churchland revolved around the issue of moral progress or moral improvement: Churchland is too optimistic about the moral community's ability to arrive at high-quality moral knowledge by the ordinary process of moral socialization that connectionism models. Does moral progress normally or typically ensue as we attempt—as sophisticated connectionist devices—to solve moral problems, reflect on the strategies we exhibit, and then, based on their success and failure, recalibrate our strategies in order to be more successful?

It might happen, indeed it sometimes does happen, that when people face new moral problems they generate new, and what is different, better moral solutions. But it is not guaranteed. It is not that connectionism fails to accurately model moral learning, it is that it does so in a way that reveals why we ought not be overly optimistic about moral progress at the societal level or at the individual level. The acquisition of socio-moral norms involves much more than the acquisition of "moral" skills. Such learning subsumes a wide assortment of competing interests. Furthermore, even when the community is focused primarily on inculcating moral norms and reliable moral response, it is typically only trying to fix them internal to itself.

In 1996, I put the worry this way:

> It is important not to think that when Churchland writes, "we have a continuing society under constant pressure to refine its categories of social and moral perception, and to modify its typical responses and expectations," the constant pressure to refine is primarily working on the moral climate. The reason we shouldn't think this is simply that there are too many interests besides moral ones vying for control of our socio-moral responses. (Flanagan 1996b, p. 208)[30]

Churchland is optimistic that persons in communities, in virtue of being learning systems with self-correcting capacities, will naturally escape local chauvinism, and that the mechanisms of learning will lead naturally to moral improvement and progress. The underlying assumption must be that moral communities are themselves progressive. The trouble is that if humans are correctly modeled by connectionist systems, we will learn what we are encouraged to learn and we will admire and avow norms that are socially endorsed. For this we will need to be well socialized. But to get beyond mere socialization we will need to understand how critical leverage is gained.

How do we learn skills that lead to genuine moral improvement as opposed to simply learning whatever norms and skills are "socially approved"? It isn't plausible to think that we normally learn how to "objectively," or to "morally" reflect on better and better moral skills and strategies based on the feedback that we get. The reason is that the skills of self-reflection we learn are themselves embedded phenomena that are generated by an array of competing interests: some moral, but some not.

How might a person, conceived as a connectionist moral learning system with self-correcting capacities, escape local chauvinism and find her way along the road to moral improvement and refinement? Set up the right way, with the right original architecture and attention to the acquisition of certain *meta-skills*, such a system can have these positive results. But we will need to supply the system with a few fancy tools such as the relevant "meta-skills" found to be necessary from other parts of psychology, social psychology in particular.

Social intuitionism predicts that, and explains why and how, different social ecologies will seek to solve their adaptive problems by tuning up or down the four modules. This explains the plural shape of socio-moral-political systems. Assume, at the time a moral code forms, that it is, in fact, a good or satisfactory solution to the adaptive problems facing the group or society.

This is too generous an assumption. The adaptive problems a society with a caste system already in place faces will be very different from those faced by one with a prior commitment to egalitarianism in place. Without getting into my reasons for saying this, if it isn't obvious, I say this much: the culture with the caste system already in place is already on the wrong foot in solving its problems unless the new adaptive challenges happen to be ones that are revealing problems with the caste system. There is the matter of power and prior power relations in determining norms.

But "make believe" for a moment that societies respond to new adaptive problems in ways that are decent design solutions to the problems. Once a moral system is in place, it is transmitted to the youth. Suppose, as seems likely, that the adaptive problems are different than those that led the elders to initiate "the original good solution." How will the youth see that this is so and adjust their socio-moral-political perspective?

A non-naturalist might seem to have an easier time with this problem than the naturalist. She can say that there is really only one adaptive problem: being good in the "eyes" of God and attaining eternal reward. When there are questions about the adequacy of a moral code, evaluate them in terms of divine standards and one will be able to set one's self straight. This is not a tactic available to the naturalist. It is predictable perhaps, but it is immature, epistemically and emotionally irresponsible. What to do if we are honest and wish not to seek meaning in mysterious stories unbefitting poetically gifted adults?

We are always, because it is the nature of the human condition, in some kind of internalist predicament. But normative assessment and adjustment need not arise internally in a narrow objectionable sense. Normative assessment, moral improvement, and so on, must come from inside the dialectical space of our own norms, the norms of those who live differently, and in the space of meta-norms for resolving disagreements or deciding to live tolerantly without agreement. This means that the internalist predicament admits of degrees and that to the degree we reflect widely, from a perspective that includes other moral conceptions we are going external in some meaningful sense of the term 'external'.

Think of the internalist predicament this way. *If* one lives within a relatively homogeneous form of life, a relatively homogeneous Space of Meaning, then there are usually only two ways normative adjustment can or will occur: internally or externally.

(a) There might be competitions in small open spaces within an arena where a norm and its rationale are settled—e.g., where it is agreed that marriage is a good and that it is forever. The small space might be one such as this: Everyone agrees about the latter, that marriage is good and it is forever, but everyone is also aware that there are some awful marital situations, so a discussion is begun about whether divorce should remain impermissible or might be permitted in rare "extreme" situations.

(b) Alternatively, a new, unexpected adaptive challenge is faced—for example, suppose that new commercial practices intrude some formerly isolated space and new issues of "economic fairness" toward members of an out-group arise. As soon as there is a sufficient level of commercial interaction between the two formerly isolated relatively homogeneous groups, they face coordination problems regarding how to do business with each other. Each group could set up independent norms governing business transactions with the other group that have no effect on their in-group dealings.[31] Or, as likely, these initially independent norms might affect and re-configure practices in and out of the group. The dialectic is still internal to norms. It always is. But it is no longer internal to *one* set of normative practice. There are *external pressures* to make internal normative adjustments. This is getting us closer to the idea that there is a legitimate conception of objective flourishing and that it involves using some sort of cross-cultural standard of wide reflective equilibrium.

My view is that to satisfy (2) we need to find reason to avow and inculcate a meta-norm—call it the *key meta-norm* (KMN)—that says this: When there is any pressing issue about norms, virtues, etc., engage in the process of wide reflective equilibrium. There are prospects both in the "relatively homogeneous" singleton cases and in cases where there is socio-normative interaction among groups that abide different norms, and especially in very heterogeneous cultures, or when globalization is in play, to develop meta-norms. But the key meta-norm is especially important. It can do real substantive work when there are pressing conflicts or confusions about what norms to avow. Furthermore it is premised on a humility that is based on accepting fallabilism.

Overall, meta-norms take many forms: they include norms that are thought to yield peace rather than war (even without normative agreement), as well as norms for "rational" discussion of what is merely conventional, makes sense, passes certain agreed-upon epistemic tests, and so on. Meta-norms insofar as they are divined and abided constitute some form of external perspective. They can be political, ethical, epistemic, or (most often) mixed. But they are not external in various all-to-familiar senses that invoke ideas of God's will or His favoritism toward some chosen people.

In 1971, when Rawls first published his *Theory of Justice*, he explicitly acknowledged making use of the method of "reflective equilibrium" as I have described it, and he claimed that *all* rational people—if placed behind the "veil of ignorance" into the "original position," where they have no idea how they will fare in the lottery of life—will choose the same two principles of justice to structure socio-moral-political life. The liberal theory of justice will be chosen because it best allows each person to develop and live in accordance with his or her own "comprehensive theory of the good."

In the early 1980s, when I began to have discussions with Rawls, he would say that the two principles of justice advanced in *Theory of Justice* would be endorsed by citizens of liberal democratic states such as America and Sweden *if* they were to go behind the veil of ignorance into the original position, and were to use the method of reflective equilibrium. What happened? How did what were initially to be the principles of justice all rational people would accept become ones that persons already living in liberal democratic states would choose?

The social intuitionist model offers an explanation, although its proponents have not given it. The straightforward way to make the point is this way: There is no such thing as universal ethical intuitions at the level Rawls was initially looking to locate them. What I mean is this: at some basic level we may get universal moral reactions. Consider Mencius's example of the child falling into a well. One may gain universal assent to "This is bad" and "I feel like doing something to stop it. Even Hitler would have." Disgust at watching innocents murdered or revulsion at gratuitous thievery might get a similar reaction. But Rawls was looking for something much more sophisticated: a universal theory of justice (which, it seemed, might be stated in terms of a principle of justice or equal liberty and "the difference principle"—that no social or economic inequality is permitted unless it helps the worst-off). His concession that the theory is one likely to be seen, espoused, or intuited by Americans and Swedes is best understood as born of the recognition that there is no method that accords with the mandatory practice of seeking wide reflective equilibrium in the sense that it is not informed by prior settled intuitions. But this fact that we are psychologically capable of going only so far in engaging in wide reflective scrutiny does mean that we cannot go external to some extent that escapes narrow, objectionably chauvinist internal reflection.

Indeed, once the latter objection is made, and the concession is made that we will not be able typically to put all our settled convictions in play when we reflect, one sees the rationale behind a familiar type of resistance to Rawls's procedure of asking folks to go behind the "veil of ignorance into the original position": Either I might object from the start to the procedure because you are in fact, asking me to put aside some of my most cherished convictions (such as my belief in the moral neutrality of capitalism or my opposition to certain kinds of sexual freedom) or I might exclaim, after I engage in the game according to your rules, "Conjurer! Trickster!" Why? Because your procedure required that I change or give up some of my most cherished moral intuitions.[32] You tricked me by making me set aside, for purposes of the exercise, my most cherished moral intuitions, and I now see that this was a big mistake.

Implications for Eudaimonics

Eudaimonics, if it is possible as a kind of empirical inquiry, must allow empirical evidence to support its conclusions. What implication, if any, does what has been said so far have for the legitimacy of the method of wide reflective equilibrium? WRE is a normative test that says we ought to test our ideas about life by bringing them into the widest space of reasons possible. The test, in order to be psychologically realizable, involves taking as genuine all credible contending options available in the Space of Meaning$^{\text{Early 21st century}}$. Or, perhaps more credible and judicious, it requires reflecting widely when there are misgivings about norms, values, and virtues and when internal scrutiny does not yield a satisfactory solution.

On the one hand, social intuitionism reveals why one might worry, as I do, that social intuitionism and connectionism (they are compatible and can blend—indeed, I recommend blending them) explain moral agreement and moral consensus but leave us poorly equipped to deal with moral conflict. Where does the naturalist go for clarity? The answer is "To the world." There is no other place or space.

Here is a thoroughly naturalistic idea that might help us understand and deal with both moral conflict and failures of moral insight. There are two findings in cognitive science that can be blended:

(1) There are well-known attribution biases that lead people to believe false things. One bias, the perseverance bias will lead people to continue thinking that they are good or bad at a task, a social intelligence task, for example, even if they are told that the data that led them to make the initial judgment were a complete fabrication. The only reliable way to get the false perseverative belief to yield is to teach the victims of the effect about the perseverance effect itself (Flanagan 1984, 1991a,b; Goldman 1986).

(2) A vast amount of recent work in cognitive and affective neuroscience has shown that bad social decisions are produced in people with disturbances in communication between emotional and cognitive systems (Damasio 1994).

Putting (1) and (2) together gives some support to the philosopher's old adage that moral insight requires hard work. If nature has gifted us with a moral system that operates mostly intuitively and if intuitions are not always reliable, then we are positioned to propose another meta-norm that instructs us to check and double-check intuitions and gut reactions: Pay close attention to your intuitive moral responses and to the confidence you experience about the validity of your norms and values. Consider alternatives.[33] Let us call this the *intuition-checking meta-norm* (ICMN).

Reason for ICMN: Evolutionary dispositions in concert with hard work at socialization on the part of the moral community are guaranteed to make you feel strong moral convictions, reinforced by your fellows who share your comprehensive view of the good. This situation works for compassionate Buddhists and creepy Nazis in pretty much the same way.[34] It will be wise to recognize this much, and thus to be willing to bring, as necessary, your moral intuitions under rational critical scrutiny. How? Engage in wide reflective equilibrium whenever the opportunity presents itself. Engaging in wide reflective equilibrium involves accepting a norm to the effect that one ought to engage in wide reflective equilibrium because it is a good idea, the best we have developed for checking virtues, norms, and rules. The method is not a guarantee of success at getting to "the right view"—remember it gives a large role to intuitions itself, but it asks that the intuitions pass certain reflective tests. If there is any method that can lead to moral justification, moral improvement, and the like, it is the one.[35]

Sometimes the right meta-norms (such as KMN and ICMN), gathered from cross-cultural reflection as well as from epistemic improvements that come from paying attention to what cognitive science teaches about our epistemic weaknesses, can engender sensitivity to false intuitive beliefs, such as that some humans are less than human. Slavery and caste systems can in theory yield to such scrutiny. Another way to gain rational leverage in intuitively well-oiled machines is this: Teach the youth both the norms abided by the elders while at the same time giving them some instruction in being watchful for internal consistencies and adaptive inadequacy.

Joshua Greene, trained first as a philosopher and now a leading researcher in "neuroethics," says this about how understanding the power of the intuitionist model—which he and Haidt (sometimes collaboratively) argue explains best the genealogy and operation of morality—can help with moral improvement:

> My interest in understanding how the moral mind/brain works is in part driven by good-old-fashioned curiosity, but I also harbor a moral, and ultimately political, agenda. As everyone knows, we humans are beset by a number of serious social problems: war, terrorism, the destruction of the environment, etc. Most people think that the cure for these ills is a heaping helping of common sense morality: "If only people everywhere would do what they know, deep down, is right, we'd all get along." I believe that the opposite is true, that the aforementioned problems are a product of well-intentioned people abiding by their respective common senses, and that the only long-run solution to these problems is for people to develop a healthy distrust of moral common sense. This is largely because our social instincts were not designed for the modern world. Nor, for that matter, were they designed to promote peace and happiness in the world for which they were designed, the world of our hunter-gatherer ancestors. My goal as a scientist, then, is to reveal our moral thinking for what it is: a complex hodgepodge of emotional responses and rational (re)constructions, shaped by biological and cultural forces, that do some things well and other things extremely poorly. My hope is that by understanding how we think, we can teach ourselves to think better, i.e. in ways that better serve the needs of humanity as a whole.[36]

My advocacy of advancing meta-norms such as KMN and ICMN is based on the same sort of goals for moral improvement, better moral, and political reasoning that Greene thinks (and I agree) are needed.

Teaching history can also help.[37] Our psychology about socio-moral norms is, ordinarily, intuitive. So we instruct the youth to watch out for intuitive discomfort, cognitive and affective dissonance, cramps in the heart-mind. Internal inconsistencies in the relevant sense might be narrow, relating pretty much to a single norm or practice (e.g., whether lying for paternalistic reasons is warranted) or might be wide and systemic (e.g., how to achieve economic justice). Take the example mentioned earlier of strict prohibitions against divorce. This is bound to create a certain amount of misery in families, perhaps in the group. One solution is to use the misery to begin discussion of allowing certain exceptions consistent with the rationale behind the prohibition. A wide, systemic inconsistency might reveal itself the way Martha Nussbaum describes. Nussbaum's work on women and human development reveals that many objectively ill, malnourished Indian widows think their health is OK; widowers, on the other hand, get that they are in poor health. One might think this is a straightforward epistemic mistake (and at a certain level it is). Hinduism, especially as practiced in villages, has all sorts of odd rules regarding the behavior and status of widows. But even without bringing in certain norms of equality from Delhi to the villages, this interesting phenomena occurs: Give both the men and women good water, decent nutrition, and the most modest health care, and now the widows and widowers both judge their health the same: not very good. The ethical point is that a small change in the objective conditions, despite sexist norms in the villages, cause the intuitive judgments of the widows to change and, in this case, to become more accurate. Finally, I could use the same example, but this time bring the villagers into normative conversation with their fellow Hindu city-dwellers. If the city dwellers can legitimately say that there is nothing internal to Hinduism (even of the form the villagers accept) that warrants the village widows to self-conceive in the way they do, then there are reasons of consistent commitment to Hindu practice for the village widows to change their low estimate of their status.

Examples of adaptive inadequacies are a dime a dozen and affect all human practices. Before Roman aqueducts, everyone wanted and needed to transport water over difficult terrain. In retrospect, most even workable ideas were not optimal, but many worked and these may have been "the best," the most satisfactory solutions available. The rise of modern industrial states with capitalistic economies brought new problems of poverty and poor care. In the seventeenth century in France and Holland there was immediate attention to the problems of the poor. These were motivated by both prudential concerns (the poor are dangerous and disgusting) and moral concerns (Jesus says care for the poor—see Gouda 1995). Both countries solved the adaptive problem by building into their respective infrastructure high tax rates, which enable strong welfare states to the present day. America, when faced with similar problems and similar possible solutions has continued to fail to solve, some would say, even address, the problem. Why don't we have universal health care? One reason is that most Americans have false views about desert and merit.

Our socialization bears too much weight from John Locke and from Horatio Alger's stories. Our system is broken and it will have to change. I am not saying that we will do so for moral reasons, although those are in play in some of the discussion. Even if the "public reasons" for changing our practices are stated as purely prudential—pertaining to public safety and good economic practices, we will eventually think that universal health care is a "good thing," something we as a society ought to provide. That's just the way things work.

At the start of this chapter, I asked whether normative mind science (in particular, eudaimonistic scientia—eudaimonics) is possible. My answer is Yes. In fact, eudaimonics is actual and has been practiced—empirically if not in the usual scientific ways—in the East and in the West for 2,500 years. I have tried to explain why the methods of eudaimonics are different, indeed why they must be different, from the methods of non-normative and observer-independent sciences. The method of wide reflective equilibrium plays an important role, perhaps the main role, in advancing moral knowledge. By combining the social intuitionist and connectionist model of morality, I showed how we can understand better how morality operates and how best, given our resources, we can deal with such matters as normative conflict and normative improvement. Finally I tried to say what a naturalist can say about ethical knowledge; in particular I tried to say something about how to think empirically about eudaimonia, about flourishing. Following Aristotle, I proposed and defended these two criteria for evaluating, respectively, subjective and objective flourishing:

(1) It is a necessary condition of subjective flourishing that the virtues an individual displays, and the norms she avows and abides, pass tests for reflective equilibrium.

(2) It is a necessary condition of objective flourishing that the virtues an individual displays, and the norms she avows and abides, pass tests for wide reflective equilibrium.

In the next chapter I do two things. First, I examine new work on the neuroscience of happiness and well-being. Eudaimonics should be able to speak about what underpins well-being with or without subjective happiness. And it does. But there are predictable problems associated with measuring happiness, due in part to disagreement about what "happiness" (especially "true happiness") even is. Second, I examine the literature on "positive illusions." Apparently, many healthy-minded people have false beliefs. These may even be partly constitutive of what makes them happy and healthy minded. Illusions, unlike delusions, are subject to modification, albeit with resistance. But even they will make epistemically conservative types nervous. Maybe the distinction between subjective and objective notions happiness gives us some leverage. Since many worry that spirituality and religion involve objectionable illusions and/or delusions, this discussion may be seen as a preliminary to my final chapter, titled Spirituality Naturalized?

Appendix

The figure below is Edmund L. Pincoffs's list of virtues canonized (in various ways) in the West (source: Pincoffs 1986). The table (titled "the 52 mental factors at a glance") is from *Abhidhammattha Sangaha* (2000).

Personality Traits

VIRTUES

NONINSTRUMENTAL

AESTHETIC		MELIORATING			MORAL	
NOBLE	CHARMING	MEDIATING	TEMPERAMENTAL	FORMAL	MANDATORY	NONMANDATORY
dignity	gracefulness	tolerance	gentleness	civility	honesty	benevolence
virility	wittiness	reasonableness	homorousness	politeness	sincerity	altruism
magnanimity	vivaciousness	tactfulness	amiability	decency	truthfulness	selflessness
serenity	imaginativeness		cheerfulness	modesty	loyalty	sensitivity
nobility	whimsicality		warmth	hospitableness	consistency	forgivingness
	liveliness		appreciativeness	unpretentiousness	reliability	helpfulness
			openness		dependability	understandingness
			even-temperedness		trustworthiness	super honesty
			noncomplainingness		nonrecklessness	super conscientiousness
			nonvindictiveness		nonnegligence	super reliability
					nonvengefulness	
					nonbelligerence	
					nonfanaticism	

INSTRUMENTAL

AGENT INSTRUMENTAL	GROUP INSTRUMENTAL
persistence	cooperativeness
courage	'practical wisdom'
alertness	the virtues of leaders and followers
carefulness	
resourcefulness	
prudence	
energy	
strength	
cool-headedness	
determination	

Ethical Variables

Universals

Contact
Feeling
Perception
Volition
One-pointedness
Life faculty
Attention

Occasionals

Initial application
Sustained application
Decision
Energy
Zest
Desire

Unwholesome Factors

Unwholesome Universals

Delusion
Shamelessness
Fearlessness of wrong
Restlessness

Unwholesome Occasionals

Greed
Wrong view
Conceit
Hatred
Envy
Avarice
Worry
Sloth
Torpor
Doubt

Beautiful Factors

Beautiful Universals

Faith
Mindfulness
Shame
Fear of wrong
Non-greed
Non-hatred
Neutrality of mind
Tranquility of mental body
Tranquility of consciousness
Lightness of mental body
Lightness of consciousness
Malleability of mental body
Malleability of consciousness
Wieldiness of mental body
Wieldiness of consciousness
Proficiency of mental body
Proficiency of consciousness
Rectitude of mental body
Rectitude of consciousness

Abstinences

Right speech
Right action
Right livelihood

Illimitables

Compassion
Appreciative joy

Non-Delusion

Wisdom faculty

5 Neuroscience, Happiness, and Positive Illusions

My favorite extravagant headline among numerous hyperbolic ones that appeared in the third week of May 2003 is "Buddhists Lead Scientists to 'Seat of Happiness.'" This headline and associated press releases, some of which implied that I had discovered the "seat of happiness," gave rise to my Warholian 15 minutes of fame. The source of the media frenzy was an article I wrote that appeared that week in the magazine *New Scientist.* "The Colour of Happiness" was the editor's choice for the title of the piece, in which I reported on two preliminary studies on one meditating monk that showed a remarkably frisky left pre-frontal cortex (LPFC), a site well correlated with reports of positive mood as well as with a certain immunity to the normal startle response.[1] I described these preliminary results as tantalizing, and I said that we were positioned to test the hypothesis that long-term meditative practice might produce happiness. This is very different, of course, from saying that studies on Buddhist practitioners had, in fact, led scientists to the Holy Grail of the "seat of happiness." But that was the conclusion that almost every radio and TV interview insisted on, despite my protestations. My job was to explain where the seat of happiness was, why *every* Buddhist practitioner had a very active one, and how the rest of us could activate it immediately. Now, several years after my initial article appeared, I take a deep breath, and ask: What is "true happiness"? What is "true happiness" like phenomenologically? How, if there is such a thing, do we measure or assess it? What are its causes and constituents? Can we judge a person to be truly happy if he or she is happy in virtue of having certain false beliefs—what psychologists call "positive illusions"?

Three Kinds of Happiness

Happiness, however one conceives it, is some sort of state or condition with causes and constituents.[2] Here are three kinds of happiness scientists study, which they and philosophers such as myself are very interested in: $\text{Happiness}^{\text{hedonic}}$, $\text{Happiness}^{\text{subjective-wellbeing}}$, and $\text{Happiness}^{\text{eudaimonistic}}$.

Hedonics

Hedonics is "the branch of psychology that deals with pleasurable or unpleasurable states of consciousness" (Kahneman 1999, p. ix). In an essay titled "Objective Happiness," the psychologist Daniel Kahneman, a Nobel Prize winner for his work on "irrationality" in economic choice, writes: "An assessment of a person's objective happiness over a period of time can be derived from a dense record of the quality of experience at each point-instant utility. Logical analysis suggests that episodes should be evaluated by the temporal integral of instant utility. Objective happiness is defined by the average of utility over a period of time." (1999, p. 3)

In *The Singularity Is Near* (2004), Ray Kurzweil says that we are just a few years away from reading our e-mail using saccades to turn on and scroll through computer screens embedded in nanobots in contact lenses or eye-glasses. So imagine a variation on the actual methods or devices that Kahneman uses and that require self-reports or button pressing every 10 seconds or so. Call my device a "hedonometer," and imagine the nanobot is embedded in glasses or contact lenses or on the eyelashes. Just as we learn to use bifocals or progressive lenses, subjects learn, i.e., are trained up, to unconsciously but reliably register by 21 unique saccadic movements instant hedonic utility on a scale from −10 to +10, with 0 neutral. This actually would not be that hard. So suppose we want to know how happy some person is over the course of the last 5 minutes, or today, or this week. We simply read off the computational measure of objective happiness over the relevant interval. Research of this sort is being done. When last I heard, Kahneman's subjects had produced on the order of a million data points.

One might worry about measures of objective happiness with the hedonometer for a host of reasons. The general thrust of objections would be that the hedonometer doesn't measure what we want to measure. Some such objections would be methodological; some would be epistemological:

(1) There is an *interference effect* caused by the need to consciously interrupt ongoing experiences. (My hedonometer helps here, since we learn to unconsciously record instant utility, but still the mind/brain is being asked to do two things at once online.)

(2) Even if the hedonometer does reliably measure average instant utility over an interval, it makes temporal assumptions about pleasurable and unpleasurable experience-intervals that are false. Specifically, the units to be assigned utility scores are often too short. Why? Because many emotionally salient experiences and episodes have longer durations than, say, 10 seconds. (Notice that the interference effect and this "longer duration" objection create a connected damaging criticism.)

(3) Experiences of "flow," of "being in the zone," are well documented to contribute to very positive experiences of what is called "subjective well-being" (SWB), but "flow experiences" are characterized by a sort of neutral conscious hedonic tone *while* they are happening (although they get high scores afterward). According to Csikszentmihalyi, "flow is what happens to and for composers, musicians, athletes, dancers, rock-climbers, mathematicians, and philosophers when what-they-are-engaged in is going well" (1996, p. 111). Flow appears to be the main constituent of "enjoyment," although "enjoyable experiences are not high in emotion while we are in the zone" (ibid.). The hedonometer is poorly equipped to capture this. (For more on "flow," see the appendix to this chapter.) That is, a score of +10 for the previous hour of playing soccer or listening to music or painting or doing philosophy or meditating doesn't (and shouldn't) produce retroactively scores of +10 for each 10-second interval over that hour. The reason is principled: each 10-second interval was not fused with conscious hedonic tone, although the overall experience was or seems to have been.

Another type of objection is *normative*. Even if the hedonometer reliably measures point-instant utility, humans retrospectively (and possibly as they are happening in real time) evaluate experiences with standards of what *matters* that override and/or recast instant utility assessments. Norms governing mattering can be defended. Suppose at the end of one day I have an average instant utility score of 7, but suppose that there were two half-hour periods during which I experienced bad lower-back pain that I scored −9 to −6, resulting in an average of −8 for that interval. During another half-hour period I had great sex with someone I love, to which I assign scores from +6 to +9, for an average of +8. (Make the sequence such that the sex occurred first so you can't complain about a recency effect.) At day's end, I produce a narrative summary of my day. I report satisfaction with the experience with my lover but don't even mention the backache. A true believer in measuring instant hedonic utility might say that this is due to a familiar bias that leads people—and thus that has led me—to spin their narratives positively (self-serving bias, SSB). Whereas the defender of measuring subjective well-being or what is different, eudaimonistic well-being, will say that the sexual episode with my lover matters to the overall quality of my life, the backache doesn't.[3] There is no mistake. This is, in fact, how the dialectic plays out.

Subjective Well-Being

Researchers who favor measuring subjective well-being over objective hedonic satisfaction think of SWB as some sort of function of life satisfaction, pleasant emotions, satisfaction with domains such as work and health, feelings of fulfillment and meaning, and low levels of negative emotions. Measuring SWB (or attempting to) has been the major approach in "positive psychology" since the 1970s, well before the field was officially given that name in the first few years of the 21st century.

Since 1972 a representative sample of Americans have been asked this question by the government as a part of the General Social Survey (GSS): "Taken all together, how would you say things are today—would you say you are very happy, pretty happy, or not too happy?" The steady finding is this: 32 percent very happy, 56 percent pretty happy, 12 percent not too happy. Almost everyone reads the relevant self-reports as involving judgments or beliefs about the subject's overall life, despite the fact that the question focuses specifically on how things are now (it asks "How would you say things are *today*?"). And almost all books on the topic use the statistics to support the claim that 80–88 percent of Americans are happy, notwithstanding the sales of anti-depressants.

Even if we think the answer to the GSS question is some sort of all-things-considered judgment about how my life has gone so far, or seems to be going now, no information is given about what I believe are the

Table 5.1

Life satisfaction of selected groups. Scores potentially range from 1 (extremely dissatisfied) to 7 (extremely satisfied). Neutral point of scale is 4.0. Source: Ed Diener; used with his permission.

Forbes richest Americans	5.8
Maasai (East African tribal people)	5.4
Pennsylvania Amish	5.1
Inughuit, northern Greenland	5.1
US college student	4.9
US cloistered nuns	4.8
Illinois nurses	4.8
Calcutta slum dwellers	4.4
Calcutta sex workers	3.6
Calcutta homeless	3.2
Mental outpatients entering therapy	2.9
California homeless	2.8
Mental inpatients (hospitalized)	2.4
Prisoners (newly jailed men in county jail)	2.4
Detroit sex workers	2.1

causal contributors to my happiness. The GSS question is very different from this question: How would you rate your satisfaction in these domains $(d_1, d_2, \ldots, d_n)$? (Here d is a variable whose subscripted tokens pick out domains that are thought to "matter" most in determining overall life satisfaction, family, friendship, work, and getting into heaven.[4]) Unlike measurements with the hedonometer, "mattering" is measured. Of course, one may be mistaken.[5] If there was good scientific evidence that my backache affects my assessment or my overall mood significantly despite my not thinking so, then I make a mistake about the nature, conditions, and causes of my state.[6]

Charts such as those reproduced here as tables 5.1–5.4 are common. One might wonder what these charts really measure. Suppose that SWB scores give information about overall life satisfaction or, what is different, satisfaction in whatever domains are picked out by psychologists. Most studies let subjects determine which domains are to be assigned values. (For an exception, see table 5.3.) But this means that, unless the psychologists are lucky, we can have no real confidence that mattering is well measured. I can make

Table 5.2
Subjective well-being scores of poor people in Calcutta, on a scale from 1 to 7, with 5 neutral. Source: Ed Diener; used with his permission.

	Life satisfaction	Housing satisfaction	Satisfaction with self	Satisfaction with family
Slum dwellers	4.4	4.1	5.5	6.1
Sex workers	3.4	4.9	4.9	5.4
Sidewalk dwellers	2.9	3.8	4.8	4.6

this point about the norms governing mattering more clearly by giving two examples[7,8]:

1. Imagine a Calcutta sex worker who rates her work as very unsatisfying and means it. But she engages in mindfulness training and learns to apply a large discount rate to her job when assigning an overall value to her SWB. Is she in the grip of an illusion, or has she simply done something adaptive with how she conceives of what matters to her overall SWB? The answer would depend, I think, on having more information than I have given. The point is that one can easily imagine that both discount and value-added strategies are deployed all the time, and that we apply them with some rationality using norms governing what matters (more or less).

2. Suppose, as is true, that I, Owen, give no value to getting to heaven. If I am not asked for an overall SWB rating, but instead one is assigned by whatever mathematical function is favored—suppose it is simply an additive average—then my score will be lower than the one I would give, since the *d*, getting to heaven (see table 5.3), isn't something I believe in, care about, etc. It doesn't matter one iota to me. In fact, I feel bad for others that it matters to them, since I consider the possibility delusional.

Eudaimonistic Happiness

Eudaimonistic happiness comes in many flavors depending on the normative conception of flourishing in play. One starts with a certain normative conception of what contributes to a truly good life and sees how individuals fare in relation to it or to each component of it. Individual subjects don't choose the domains or the weights assigned to them. The normative

Table 5.3
College student respondents' importance ratings of happiness and other values. Potential responses range from 9 (extremely important) to 1 (not at all important). Nations shown are selected from 47 nations sampled. Source: Ed Diener; used with his permission.

	Happiness	Wealth	Love	Health	Meaning	Attractiveness	Getting to Heaven
Overall	8.0	6.8	7.9	7.9	7.3	6.3	6.7
Brazil	8.7	6.9	8.7	8.6	8.3	6.4	7.8
Canada	8.6	7.1	8.6	8.2	8.1	6.3	6.5
Chile	8.6	6.9	8.6	8.1	8.2	5.8	7.8
Portugal	8.6	6.5	8.8	8.6	8.4	5.6	6.4
Australia	8.3	6.5	8.2	7.9	7.6	5.9	6.8
Nepal	8.2	7.2	8.4	8.3	8.3	6.3	7.2
Egypt	8.1	7.6	7.4	8.0	7.1	7.2	7.3
Ghana	8.1	7.1	7.9	8.0	7.2	6.7	8.1
Greece	8.1	6.8	8.3	8.8	8.2	6.1	6.4
Nigeria	8.1	7.4	8.2	8.4	7.1	7.2	8.4
Russia	8.1	7.3	7.9	8.2	7.9	6.9	5.9
Thailand	8.1	7.4	7.4	7.8	7.6	6.4	6.8
United States	8.1	6.7	8.3	7.6	7.6	6.2	7.3
Indonesia	8.0	7.2	7.9	8.0	7.9	6.7	8.2
Poland	8.0	6.8	8.2	8.2	8.1	6.2	7.5
Iran	7.8	7.0	8.1	8.5	7.9	6.6	7.9
South Africa	7.8	6.4	7.6	7.6	7.2	5.5	8.2
Bangladesh	7.7	6.7	7.3	6.6	6.6	6.9	6.8
Germany	7.7	6.8	8.6	8.4	8.0	6.5	5.7
Switzerland	7.6	6.4	8.4	8.3	8.0	6.2	6.1
India	7.5	7.0	7.5	7.8	7.5	5.7	6.6
Kuwait	7.4	7.3	7.8	8.4	7.9	8.1	7.9
Japan	7.4	6.6	7.8	7.8	6.8	5.9	6.1
Uganda	7.4	6.7	8.4	7.4	7.9	7.1	8.0
China	7.3	7.0	7.4	7.8	7.5	6.1	5.0
Malaysia	7.3	6.6	7.2	7.4	7.2	6.5	7.2

Table 5.4

Subjective well-being values for various nations, on a scale of 1 to 9. Source: World Values Study Group (1994).

	Life satisfaction	Hedonic balance	Positive affect	Negative affect
Bulgaria	5.03	0.91	1.93	1.01
Russia	5.37	0.29	1.69	1.41
Belarus	5.52	0.77	2.12	1.35
Latvia	5.70	0.92	2.00	1.08
Romania	5.88	0.71	2.34	1.63
Estonia	6.00	0.76	2.06	1.28
Lithuania	6.01	0.60	1.86	1.26
Hungary	6.03	0.85	1.96	1.11
India	6.21	0.33	1.41	1.09
South Africa	6.22	1.15	2.59	1.44
Slovenia	6.29	1.53	2.33	0.80
Czech Republic	6.30	0.76	1.84	1.08
Nigeria	6.40	1.56	2.92	1.36
Turkey	6.41	0.59	3.09	2.50
Japan	6.53	0.39	1.12	0.72
Poland	6.64	1.24	2.45	1.21
South Korea	6.69	—	—	—
East Germany	6.72	1.25	3.05	1.80
France	6.76	1.33	2.34	1.01
China	7.05	1.26	2.34	1.08
Portugal	7.10	1.33	2.27	0.94
Spain	7.13	0.70	1.59	0.89
West Germany	7.22	1.43	3.23	1.79
Italy	7.24	1.21	2.04	0.84
Argentina	7.25	1.26	2.45	1.19
Brazil	7.39	1.18	2.85	1.68
Mexico	7.41	1.38	2.68	1.30
Britain	7.48	1.64	2.89	1.25
Chile	7.55	1.03	2.78	1.75
Belgium	7.67	1.54	2.46	0.93
Finland	7.68	1.18	2.33	1.15
Norway	7.68	1.59	2.54	0.95
United States	7.71	2.21	3.49	1.27
Austria	7.74	1.77	2.90	1.13
Netherlands	7.84	1.81	2.91	1.10
Ireland	7.87	1.99	2.89	0.90
Canada	7.88	2.31	3.47	1.15
Sweden	7.97	2.90	3.63	0.73
Iceland	8.02	2.50	3.29	0.78
Denmark	8.16	1.90	2.83	0.93
Switzerland	8.39	1.14	1.39	0.24

conception does all that work. Measures of eudaimonistic well-being (EWB) incorporate a large objective component that is different from the objectivity of instant utility measures. EWB measures could be wildly divergent from SWB scores, as well as from measures of average instant hedonic utility.

I will use an interesting idea from Aristotle to illustrate this point. Suppose a woman dies having lived what subjectively seemed like an excellent life (high SWB) and that is so judged by others. At her memorial service this is said. Is this dead person eudaimon? Since she is dead, one who favors subjective evaluation might say that she *was* eudaimon. But since she is dead and no longer a subject of experience, she isn't happy or unhappy *now*. Aristotle could agree, but would still say this: We can't even judge whether she *was* eudaimon, even supposing she and we have excellent information about her character and her life, until we see how her children and grandchildren turn out! Even if we believe that Athenians must have had much more causal power over their kids than we do, Aristotle's point still holds interest. The quality of a person's life, whether she was or is eudaimon, is not simply a matter of how she feels about herself, nor about how others judge her; it depends in addition on her causal contribution to the quality of the character and lives of the other persons she is at least partly responsible for creating. Karma!

One might think that Aristotle is being excessively moralistic, especially if we assume (a) that he overestimates the causal power of parents and grandparents to affect the character of their charges and (b) that if a person has done her best to build good character for herself and in her charges, she has done all we can reasonably expect. This seems right. But the point I want to draw attention to, which the illustration reveals, is the divide between subjective and objective evaluation of eudaimonia. (It was this thread in Aristotle that allowed me, in chapter 4, to distinguish subjective and objective conceptions of flourishing.) Not only is it a consequence of Aristotle's view that the person I depicted cannot be judged eudaimon until certain later objective states of affairs are assessed, so too are these two sorts of possible cases:

(1) A person who does not feel very happy (i.e., who has low SWB) is judged by her compatriots as unusual in virtue of being loving and compassionate even when being this way doesn't seem to serve her self-interest. At best, she is thought to be a sweet naïf. Suppose her children

and grandchildren turn out just like her and are judged the same as she was by their compatriots, as sweet but on the fringe. Is she, or was she, eudaimon? Yes. It is too bad she never felt the self-esteem she deserved, but she was eudaimon. Her society was simply a poor judge of goodness, and she, not surprisingly, was affected in how she felt about herself by how others saw her. Ergo, her low self-esteem.[9]

(2) Imagine that a greedy entrepreneur feels very good about himself, his family, and his work. He obeys the laws, national and international, but will do whatever is best for his business. He is off the charts SWB-wise, and his hedonometer keeps breaking from overuse of the +10 key. Is he eudaimon? It's doubtful. I can't help but comment that people such as these are abundant in business and in politics. Social endorsement of capitalistic economic practices allows such people to think they are good.

Carol Ryff and Corey Keyes (Ryff 1989; Ryff and Keyes 1995) have recommended a scheme to measure *psychological well-being* (PWB). PWB is different from SWB, falling more into this eudaimonistic category. PWB is a way of evaluating well-being that takes into account the extent to which subjects *endorse* such values as high levels of autonomy, environmental mastery, personal growth, positive relations, living purposefully and meaningfully, and of living in a way that supports positive self-regard. One can see the difference between SWB and PWB by remembering that on one standard way of measuring SWB scores they are determined as a function of four possibly separable components: life satisfaction, satisfaction with specific domains (e.g., work and family), amount and/or frequency and/or tone of positive affective experience and negative affective experience (where in some cases negative experience is subtracted from positive experience to compute overall hedonic effect). PWB starts with a conception of worthy aspirations and then measures whether and how much individuals endorse these aspirations and values and, in principle, how well they see themselves doing in achieving them.[10]

These three ways of conceiving and measuring happiness are woven throughout subsequent discussion (even when they are not directly mentioned), so keep them in mind. Out of fairness I should say that there is variable sensitivity among researchers who work in positive psychology about the problems and pitfalls of different measurement tools. One strategy that, in certain respects, is already in use is to keep using different measurement tools, objective hedonic measurement, SWB, and eudaimonistic ones, such as PWB, and let the dialectic play out among them, examining what the advantages and disadvantages of each are as research progresses.

The Meditating Monk and Neurophenomenology

In my 1992 book *Consciousness Reconsidered*,[11] I proposed use of a method for studying the mind/brain. I called it "the natural method" and recommended it as the right method to adopt if one is a naturalist about mind. The basic idea is to triangulate a subject domain by coordinating phenomenological data with psychological and behavioral data and both with neuroscientific data. Francisco Varela (1999a) called his almost identical method "neurophenomenology."[12] That is a sexier, catchier name, so I will use it. The work on the meditating monk is an instructive example of how profitable the method can be. Since more is involved in happiness, positive mood, well-being, and eudaimonia than what is going on in the brain, the task is eventually, or even at the same time, to make surmises about which causes and condition in genes, in fetal development, in upbringing, in education, in moral and spiritual commitments, and in the social world generally are correlated with various brain indices of well-being.

Starting in the early 1970s, three young graduate students met in Cambridge, Massachusetts. Richard Davidson and Daniel Goleman were Ph.D. students in psychology at Harvard; Jon Kabat-Zinn was a Ph.D. student in molecular biology at MIT. They met because they were all interested in the practice of meditation. Very quickly, Davidson and Goleman set to publishing some work on the relation of pre-frontal cortical activity and positive and negative affect, at least one on the connection of meditation and positive affect (Davidson, Goleman, and Schwartz 1976).[13] Before their studies, there were some good data indicating that first-person SWB reports linked to differential activity across pre-frontal cortex (PFC). Leftward activity (LPFC) was correlated with positive affect; rightward activity (RPFC) with negative affect. One background theory that might explain this hemispheric difference, and that has received increased support over the years is this: PFC is important, maybe the most important area, when it comes to hatching plans and executing action, either approaching a goal or withdrawing from an aversive situation that is undesirable. LPFC is more active when we are going after what we want. Activity in RPFC increases when we want to escape what is undesirable.[14]

The pre-frontal lobes are relatively recently evolved structures (in ancestors of *Homo sapiens*) and have, as I have said, long been known to be important in foresight, planning, and self-control. The confirmation of the fact that pre-frontal cortices are also crucially implicated in emotion, mood, and temperament is exciting because it lends some insight into *one* area where a well-functioning mind coordinates cognition, mood, and emotion. How exactly the coordination is accomplished is something about which little is known at this time.[15]

In any case, 30 years after Davidson, Goleman, and Kabat-Zinn's first foray into studying the relation between PFC and meditation was first done, and thus at the time I wrote "The Colour of Happiness" (2003), this view of the role(s) of PFC was fairly well entrenched. I surmise, based on the science, that it, or something in its vicinity, is true.

In 2000, Matthieu Ricard, an experienced Buddhist monk (born and raised in France), was brought to Madison, Wisconsin, where Davidson's Center for Affective Neuroscience is housed, and his brain was studied. Lo and behold, Matthieu's left pre-frontal cortex lit up brightly (thus the editor's choice of the word 'colour' for the title of my article). Indeed, his left side lit up brightly and more leftward than that of any other individual tested in previous studies (approximately 175 subjects). It was literally off the charts. However, none of the earlier studies involved people meditating while the scanning was underway. (In the meditating monk's case, most of the meditation was on compassion and loving-kindness.) Scientific problems related to different tasks performed by Matthieu and the previous 175 subjects did not prevent various media sources from announcing that scientists had established that Buddhist meditation produces "true" happiness. Indeed, I was asked repeatedly whether Matthieu was the happiest person ever studied, and even whether he was the happiest person on Earth. One could imagine this line of conversation culminating in the question "Is Matthieu the happiest person ever to exist?" (If that happens, I will say "Yes.")

If you are even slightly tempted to let your thoughts roam in this direction, stop and consider these obvious facts: Being "leftmost" PFC-wise is not in any way like "being the tallest." Whoever is the tallest is the tallest, but whoever is the leftmost, despite being the leftmost, isn't the happiest, no matter what standard of happiness one is using. The consensus in neuroscience is that, for most and perhaps all complex mental states, individual brains "do the same thing" in sometimes very different ways, and at somewhat different locales. Suppose two people both think "that patch is red" in response to the exact same red patch stimulus. Assume that both are "having the exact same thought," although it must be said even this assumption is controversial. We might after all experience red a bit differently, perception of red things might cause different associations, and so on. Bracket these worries. Assume that whatever else goes on when each of these two individuals think "that is a red path," both think that much, and each thinks the thought in the same way as far as that red patch goes. If so there will be brain activation in each individual that is *that* thought or is the neural correlate of *that* thought. But no one expects two different brains to have exactly the same thought (assuming they are) in a way that is subserved by perfectly identical neural activation. The consensus is that the exact same thought can be realized (indeed is likely to be realized) in different brains in somewhat different ways. The same for phenomenologically identical or very similar emotional states.[16]

For all we currently know, the subject who tests 25th or 35th or even 70th from the leftmost point so far plotted might be, according to all the evidence taken together—phenomenological, behavioral, hormonal, neurochemical—the happiest person ever tested.[17] Left-side pre-frontal activity is a reliable measure of positive affect, but no one has asserted let alone confirmed that among the group of "lefties," the leftmost individual is the happiest. One problem is due to the fact that the concepts of "positive mood" and "affect," even more so "happiness," are not fine-grained enough, nor sufficiently well operationalized by the scientists who use them, so that we know what *specific* kind of positive mood or emotional state is attached to a lit-up area.[18]

The important point is that for all anyone knows at this point, a happy life whose source is family might light up the brain in the same way as a happy life whose source is virtue or even money. One can control for this of course by being careful about the phenomenological and psychological-behavioral aspects of the *neurophenomenology*. Suppose that Donald Trump's, Rupert Murdoch's, and Hugh Hefner's LPFC light up just as Matthieu's did, with the ratio of LPFC to RPFC activity the same. We know that the causes and constituents of their "happiness" are different from Matthieu's and thus, it seems that one ought to say that the happiness itself is different in kind. That is, even though the brain at this level of analysis doesn't reveal what the differences are, there are big differences. From a naturalistic perspective, these had better show up, at least as encoded, somewhere.

So we say that despite substantial LPFC profiles among this foursome, both the causes of their happiness are different, and the contents of their happiness (what there happiness is about) are different. Let me explain what I mean by "about-ness." Philosophical talk of the aboutness of mental states originates in what Franz Brentano (adopting ideas from Thomas Aquinas) called "intentionality" (from the Latin *intendo* = to aim at). Feeling yucky might not be about anything, but believing [snow is white] is *about* the color of snow. So even if A is happy to +9 *about* the fact that [she is very rich] and B is happy to +9 about the fact that she [directs an AIDS orphanage], they are happy in different ways. The *content* of their happiness is different even if the *degree* of their happiness is the same. There are no brain imaging or scanning techniques that (as of now) distinguish among contents. All information about *content* needs to come from first-person and third-person narratives and the like.

This much shows that a state of happiness is not to be typed or classified solely by neural markers in PFC. This is where other measurement tools come in handy since SWB ratings, if fine-grained enough, will show obvious differences between what matters to Hugh Hefner and what matters to Matthieu. And mattering and what I call content are related in a way I will leave unspecified, but that is worth working out precisely.

In any case, 'happiness' is a polysemic term; it has many different meanings. For this reason, when discussing the topic I almost always revert to superscripting, so there is happiness^{Buddha}, happiness^{Aristotle}, happiness$^{\text{Local hedonist club}}$, happiness$^{\text{Marquis de sade}}$, happiness$^{\text{standard American}}$ = happy$^{\text{joy-joy-click-your-heels}}$. Aristotle pointed out that everyone says eudaimonia is the greatest good while meaning different things by the term. The same situation obtains today. Even those who say that happiness is what they want more than anything else may well mean different things by 'happiness' or, what is different, may overrate happiness colloquially understood as happiness$^{\text{standard American}}$ = happy$^{\text{joy-joy-click-your-heels}}$.

The point about overrating happiness entails that even if one goes along with the superscripting strategy it doesn't remove all the problems. Individuals might legitimately claim that, even though they judge their lives as very meaningful as, say, an Aristotelian, or a Buddhist, or a Christian, feeling happy is not even an issue. And thus ascribing $\text{happiness}^{\text{Aristotle}}$ or $\text{happiness}^{\text{Buddha}}$ misses the point.

Was the Buddha happy? Was Jesus happy? Was Confucius (Kongzi) happy? The answers do not seem obviously Yes according to common contemporary usage of the term 'happy'. When I ask my students these questions they are unsure what to say. Many will say it doesn't matter or that is not the point. They are on to something.

On the other hand, no one who seeks $\text{happiness}^{\text{Local hedonist club}}$, $\text{happiness}^{\text{Marquis de Sade}}$, or $\text{happiness}^{\text{standard American}} = \text{happy}^{\text{joy-joy-click-your-heels}}$ will say that happiness is not the issue, not what matters. It matters to them. In the Buddhist case, equanimity is one of the four noble illimitables (the other three are compassion, loving-kindness, and appreciative or sympathetic joy—being happy for the successes of others). 'Equanimity' describes an abiding condition of "heart/mind" (this is the right translation for all Asian terms that we might wish to translate as 'mind'). Equanimity, whatever exactly it names, might be said (like flow) to be intermediate between anxiety and boredom. It might in certain cases feel blissful, but it might not. As far as my own linguistic intuitions go, words like 'equanimity' and 'serenity' are not, even in American English, in the same family as $\text{happiness}^{\text{standard American}} = \text{happy}^{\text{joy-joy-click-your-heels}}$.

In any case, as I have indicated, before the study of the meditating monk there had been a number of excellent studies on positive affect and the brain (Davidson and Irwin 1999; Davidson 2000; Davidson et al. 2002; Davidson and Hugdahl 2002). This research showed that there was a distinction between basal (baseline) and tonic (variable) reports of "well-being" or "positive mood" as reported first-personally and as measured in PFC. The still-dominant view is that each person has a characteristic baseline ratio of LPFC:RPFC activity, and then various experiences (or stimuli) result in changes (which some say "overlay" the basal condition).[19] Thus when subjects are shown pleasant pictures (say, sunsets), scans (PET or fMRI) or skull measurements of activity (EEG), reveal increased left side activity in pre-frontal cortex.[20] Whereas when subjects see unpleasant pictures (say, a human cadaver), activity moves rightward. Furthermore, people who report themselves generally to be happy, upbeat, and the like, show more stable left side activity than individuals who report feeling sad or depressed in whom the right side of pre-frontal cortex is more active.

Positive mood has two faces. Subjectively, phenomenologically, or first-personally, it reveals itself in a way that an individual feels, and about which she typically can report on (although subjects commonly report difficulty describing exactly what the positive state is like). Objectively, the subjective feeling state is reliably correlated with a high degree of leftward pre-frontal activity. Thus we can say that *if* a subject is experiencing happiness or, what is possibly different, is in a good mood, then left pre-frontal cortex is or gets frisky, or bright, or even colorful depending on whether you use EEG, fMRI, or PET.

Most people (70–80 percent) report themselves to be happy as opposed to not so happy, sad, or depressed. Davidson found that in a normal population (of undergraduates) pre-frontal lobe activity is distributed in a fashion that corresponds to the phenomenology, a left-leaning hyperbola. Those who are very happy are fewer than those who are pretty happy, but both groups are larger than *les miserables*.

I take it that any finding to the effect that Buddhist practitioners are happier than most would be a statistical finding that significantly more than, say 25 percent are in the first, very happy group. Since the neurophenomenological curve reveals that in a normal undergraduate population approximately 18 percent fall into the "very happy" group, then a finding that 25–30 percent in that group for a representative sample of Buddhist practitioners would be statistically astounding. A somewhat lower percentage would still be impressive. As I write the data don't exist but studies are underway. The Wisconsin group has a big research program underway studying adepts, and Alan Wallace and his colleagues in Santa Barbara are undertaking a long term study of shamatha meditation (training in sustained voluntary attention) and its effects on happiness.

There are many people watching this research closely. And most of those I know are betting that the greater happiness hypothesis will be confirmed. I'm not sure. Let me explain why.

Matthieu radiates happiness. I am not sure how much he values happiness, but I know he values virtue and wisdom more. But he is, there is no doubt about, a very happy man. Most of the Buddhists I know, however, will say that happiness is *not,* in almost all its usual senses, what matters. Wisdom and virtue are what matters and being wise and virtuous typically bring equanimity. The next time I see Matthieu, I will ask him what he thinks. I predict he will say this too, despite the fact that he is very happy—by my standards, at least.[21]

So Matthieu is very happy and equaniminous, and he is "off the charts" LPFC-wise. But he is just one individual. What should we expect in studying other adepts, Buddhist monks who have practiced meditation for many years or secular Western practitioners of mindfulness practice? I suspect we will find this: subjectively if we listen carefully we will hear them reporting that they are happy (if they are willing to use the word at all) = serene and equaninimous, not happiness$^{\text{standard American}}$ = happy$^{\text{joy-joy-click-your-heels}}$. One reason for my confidence is saying this is that I know enough about Buddhism to know that if there is such a thing as happiness^{Buddha}, that Buddhists would endorse as desirable, it will involve two aspects (Flanagan 2006a):

1. a stable sense of serenity and contentment (*not* the sort of happy-happy/joy-joy/click-your-heels feeling state that is widely sought and promoted in the West as the best kind of happiness)
2. that this serene and contented state is caused or constituted by enlightenment or wisdom *and* virtue or goodness[22] as these are characterized within Buddhist philosophy.[23]

Regarding the current state of research, there are in fact *no* scientific studies yet on Buddhism as a lived philosophy and spiritual tradition, in any of its forms, and happiness. What we do have are a few scientific studies that involve examining meditators—mostly experienced Tibetan Mahayana practitioners from France, America, and Northern India (Lutz, et al. 2004), or individuals new to the practice of Zen and mindfulness meditation (Kabat-Zinn 1995; Davidson and Kabat-Zinn et al. 2003; Rosenkrantz et al. 2003). What has been found is interesting and important:

- Mindfulness practices lower stress and cause relaxation. Cortisol flow, the "natural killer" is contained, moderated, regulated. This helps with health and longevity.
- Mindfulness practice increase the number of influenza antibodies in meditators relative to controls both of whom have been given flu shots (Davidson and Kabat-Zinn 2003; Rosenkrantz et al. 2003).

- Mindfulness practice among adepts produces widespread synchronized gamma oscillations. This is rare in subjects who do not practice mindfulness, but such activity bespeaks of a mind/brain that is active, attentive, and very well focused.[24]

- Adepts doing compassion meditation report increased intensity in a way that maps nicely onto increasing gamma activity and increased global synchrony of gamma.

- Matthieu is excellent—better than anyone ever tested—at controlling his startle response. The bets are that other very experienced adepts, as well as experienced mindfulness practitioners, will be able to do so as well.

- Matthieu and three other experienced meditators are "off the charts" when it comes to reading facial "micro-expressions."[25]

The findings just mentioned have *nothing* directly to do with measuring happiness, and in this way are utterly different in kind from the PFC work. There is one exception: Davidson and Kabat-Zinn reported mood improvements as well as increased influenza antibodies. Nonetheless, all these studies show the worth of certain mindfulness practices. Supposing that researchers return to their initial devotion to testing for "happiness," and assuming that we eventually succeed at measuring the effects of different types of practice on happiness, we need to be extremely clear about what sort of happiness, if any, the practice aims at or promises. This requirement is a general one for doing good science in this area. If one wants to study ordinary Americans, Aristotelians, utilitarians, Trappist monks, secular humanists, scientific naturalists, or members of the local chapter of the "Hedonist Club," one will want their experts to specify the kind of happiness they claim to seek or to achieve. And one will want information on what aspects of their form of life they think lead to attaining their theory-specific form of happiness.[26] Only with such information at our disposal can scientists construct experiments to "look and see" if the kind of happiness sought is attained and whether the practices thought to produce that kind of happiness are causally implicated in its production. Such work will involve a mother lode of careful anthropological and sociological analysis as well as psychology and neuroscience.[27,28]

We also need much more philosophical work in which different theory and tradition specific conceptions of "true happiness" receive articulation.

We know that Aristotle, Epicurus, Buddha, Confucius, Mencius, Jesus, and Mohammed all put forward somewhat different philosophical conceptions of an excellent human life with somewhat different conceptions of what constitutes true happiness. With these different conceptions well articulated, we can look at brain activity within and across advocates and practitioners of different traditions to see what similarities and differences our mappings reveal. The same strategy should work for negative emotions and destructive mental states. Get well-honed first-person reports from subjects on the negative states they experience and then look for brain correlates. With such data in hand we can then test Buddhist techniques, say, meditation on compassion, which are thought to provide antidotes for anger, hatred, and avarice. Along with first-person reports on any experienced change in mood or emotion we can look and see what, if anything, reconfigures itself brain-wise. We can do the same for practices from other traditions. Eventually, we will want to coordinate such studies with the ever-deeper knowledge of the connections among virtue, mental health, well-being, and human flourishing, allowing science and philosophy to speak together about what practices seem best suited to make for truly rich and meaningful lives. At this distant point, with an array of conceptions of excellent human lives before us, as well as deep knowledge of how the brains of devotees of these different traditions look and work, we should be able to speak much more clearly about the nature(s) and types of happiness and flourishing than we can now.

The more theory-specific conceptions of virtue, well-being, and flourishing that we have, so much the better will our understanding be of the constituents of happiness, if, that is, "happiness" stays on our radar as what matters most all-things-considered. Overlapping consensus on the components of these things will, no doubt, reveal itself. Differences in conceptions of virtue, well-being, and flourishing will also reveal themselves. The overlaps and the differences can be discussed and debated at the philosophical level from a normative ethical perspective, and the scientists can chime in, wearing philosophical hats if they wish, but equally important, telling us how the brains of practitioners from different traditions light up, which neurochemicals rise and fall, and so on.

Inter-theoretical conversation such as I am envisioning will put us in the exciting position of being able (a) to have a better idea of the fine-grained states we are looking for and (b) to compare different theories in terms of the goods they claim to produce and hopefully do, in fact, produce.

For those of us who are convinced that Buddhism is a noble path to wisdom, virtue, and happiness^{Buddha}, and especially at this time when some scientists claim to be reaching pay-dirt in the empirical exploration and confirmation of what many Buddhist practitioners already claim to know, it is necessary to speak with maximal precision about what practices, Buddhist and others, are thought to produce what sorts of specific positive states of mind and body. Overall, this sort of inquiry provides a truly exciting, unique, and heretofore unimagined opportunity for mind scientists, practitioners, and philosophers from different traditions to join together in a conversation that combines time-tested noble ideals with new-fangled gadgetry to understand ourselves more deeply and to live well, better than we do now.

Positive Illusions

False beliefs abound. It would be worrisome if there was a tendency to hold false positive beliefs about oneself and if, across social environments, doing so causally contributed to subjective happiness. It may be an idle wish of eudaimonics, as I conceive it, that we be able to explain that if there is any such tendency to harbor positive illusions that we would flourish more if we learned to overcome it as far as possible. I have expressed hope that what I call our platonic orientation to locate what is true, what is good, and what is beautiful can succeed, that the true, the good, and the beautiful can live together harmoniously, and that their doing so is a condition for genuine flourishing.

Western epistemology is built around the idea that we ought not have false beliefs (after childhood, at least). In the East, Buddhism and Confucianism say pretty much the same. In Buddhism, *moha*—false belief—is one of the three poisons (basic noxious tendencies of persons) that can pretty much ruin a life. '*Moha*' is often translated as "delusion."[29]

Psychologists typically distinguish illusions and delusions this way: illusions are subject to feedback. I believe that [*I* will *never* get prostate cancer] even though I am told that my chances are 1 in 6 in virtue of being male. I get prostate cancer, then I give up the illusion that I won't. "Delusions" do not similarly yield to new information. It would be an odd and very disturbed duck who believed he would never get prostate cancer, did and insisted that he didn't have it. This would be delusional.

Until I first visited Italy, when I was 21, I believed that spaghetti was a rice-like crop harvested in Italy. I learned this was false. No harm was done. My false belief simply yielded to the truth.

The sorts of "positive illusions" I aim to discuss are epistemically worrisome. They are unlike ordinary false beliefs (my spaghetti belief was an ordinary one) because they are strongly resistant to disconfirmation. Delusions are often totally immune from disconfirmation. Why are "positive illusions" a problem? Because, according to some psychologists they contribute to happiness, well-being, optimism, and so on. Indeed, according to some studies, the only people who lack such illusions—and see things as they are—are "moderately depressed!"

I recently came across a passage in which Aristotle, Plato's most excellent student (and Plato was a hater of false beliefs if ever there was one), seems to endorse positive illusions. Aristotle says: "We ought not to follow the proverb-writers, and 'think human, since you are human,' or 'think mortal, since you are mortal.' Rather, as far as we can go, we ought to be proimmortal, and go to all lengths to live a life that expresses our supreme element; for however much this element may lack in bulk, by much more it surpasses everything in power and value." (X: 13.37) Earlier in the *Nicomachean Ethics* Aristotle had stated that a life of pure contemplation is the best life, but that only a god can live such a life. Here, in book X, he seems to endorse the contemplative life as a realistic human aspiration and at the same time to think that one thing worth contemplating is positive illusions. Why might this be a good idea? Perhaps Aristotle thinks that it is our capacity to put ourselves into an overreaching frame of mind that allows us to actually overachieve what we normally can't achieve. Could overachieving—morally, athletically, etc.—normally involve setting impossible goals for oneself?

I argued in chapters 1 and 2 that humans in virtue of our nature possess some sort of platonic orientation: When conditions allow we like to live in the vicinity of what is "true," "good," and "beautiful." I was careful not to endorse Plato's actual view that this is due to an impulse to return to the world of the Forms (*Eidos*) where our souls lived before birth and for which we now as embodied beings recollect (dimly) and thus long for. No, I situated the platonic orientation in our nature as social animals as depicted by neo-Darwinism. I should add that Plato, despite thinking we are born with the relevant orientation, did not think it was easy to locate, (re-) discover, and embody the true, the good, and the beautiful.

Most people after all live in the famous cave and it is only through the sort of education and training endorsed in the *Republic* and embodied in Plato's Academy where we, possibly only some few, learn to know reality deeply and truthfully. It is good to keep these two issues separate: knowing what is true and knowing deep truths.

What if the actual way or degree we are endowed with the platonic orientation is weak, and that, in fact, our aspirations to live well, in the vicinity of what is beautiful and worthy, and to be happy, require turning down our truth detectors, at least when it comes to accurate self-assessment and assessment of how those we care about are faring? The idea would be that there is something like competition among the "forms." The true, the good, and the beautiful are not designed to live in perfect harmony. How bad would this be? Is there good evidence that living among certain positive illusions is required to live a happy life? I first discussed this worry in *Varieties of Moral Personality* (1991). The state of research has not changed very much. I am going to orient the present discussion toward the findings discussed in an important paper in the psychological literature: "Illusion and Well-Being," by Shelley Taylor and Jonathan Brown (1988).[30] That paper deals primarily with the relation between a certain kind of epistemic realism and mental health, but it connects this relation in various tantalizing ways with contentment and goodness.

Let me refer to the view that accurate appraisal of one's self, of one's compatriots, and of one's world are essential components of mental health as the *traditional view* (TV). The mentally healthy person has close contact with reality. She sees things for what they are even when that is not as she wishes things to be.

In a distillation of dominant views in the late 1950s, Marie Jahoda (1958) claimed that this expectation of accurate appraisal is a central component in all extant models of the mentally healthy person—for example, the models of Allport (1943), Erikson (1950), Menninger (1930), and Fromm (1955). "The perception of reality," Jahoda writes, "is called mentally healthy when what the individual sees *corresponds to what is actually there*" (1958, p. 6). "Mentally healthy perception," she wrote in an earlier paper, "means a process of viewing the world so that one is able to take in matters one wishes were different without distilling them to fit these wishes" (1953, p. 349).

TV must of course be interpreted as allowing room for persons who, through lack of interest, attention, education, or intelligence, have no views, or implausible or irrational views, on matters such as probability distributions, life after death, the geopolitical and economic scene, and human psychology. TV allows considerable latitude in the accuracy of our appraisals of the self, others, and the world based on such factors as interest and attention, as well as knowledge of "theory." Thus, the requirement that what a mentally healthy individual sees "corresponds to what is actually there" involves a somewhat loose and culturally relative sense of correspondence. That said, at least in our culture, among accurate appraisal of the self, others, and the world, the first—the requirement that our sense of self "correspond to what is actually there"—is considered normatively important. We are, it seems to me, heirs and heiresses to some version of Socrates' idea that the unexamined life is not worth living, and thus that some work at self-knowledge is a personal, social, and moral good. Maybe it is better to say that, in deploying certain norms of common-sense savvy, we expect people to know what they did, what others did, and how, and at whatever level is necessary for getting around, what is happening in the external world. Minimally in the intrapersonal case, we expect people to be in touch with their beliefs and desires as they arise, and to know and remember what they did. Socratic norms of self-knowledge go deeper and involve ethical scrutiny of our motives, aims, intentions, plans, and so on, as well as accurate self-appraisal, of seeing who we are, what we are like, and so on. I am not sure how deep we expect such knowledge to go. In part, there are certain accessions to realism: After all, we know—even without importing an orthodox version of Freudian psychology or deep knowledge of the "cognitive unconscious"—that we cannot just do a "look see" and discover who we are or what we are like, nor are our motives transparently visible for inspection, even less so what caused our motives to be as they are. This kind of knowledge of oneself requires more than a little theory. Because over world-historical time people have thought that "knowing thyself" more deeply than we are naturally disposed or equipped to matters, we have constructed psychological theories and methods to better enable us to do so. This is relevant in the present case because if we do judge that people are prone to insensitivity when it comes to self-knowledge or, what is different, that we are commonly self-deceived, then it might be that pointing this out, and explaining why, when, and how we cause ourselves to be insensitive or to believe falsely, will extinguish the effect if the cognitive-affective bias is judged first and third-personally to be undesirable. This is well known to happen with the "perseverance effect." Learn about the tendency to persevere in attaching importance to what you have learned is false feedback and you can stop doing so. If your self-esteem still suffers because someone deliberately and without ground tried to make you feel lousy, understand that this is not surprising; it is normal. With this information about how the "perseverence effect" works, it often yields (Goldman 1986).

We are imperfect information processors, but we can learn about what sorts of false or incorrectly interpreted information lead to incorrect self-assessments (negative or positive) and work to get over the tendency. Because the "perseverance effect" can be thought of as a type of bad epistemic tendency and can be overcome, that fact is directly relevant to other alleged reasoning weaknesses, such as alleged tendencies to adopt "positive illusions." But this strategy of overcoming positive illusions, assuming they are ubiquitous, is not a good idea if such illusions are essential ingredients for mental health, as Taylor and Brown claim. But are positive illusions really essential ingredients of mental health, and, what is different, of a happy mind? I think not.

Based on a meta-analysis of a wide array of studies, Taylor and Brown claim that (1) unrealistic positive self-evaluations, (2) exaggerated perceptions of control or mastery, and (3) unrealistic optimism are (a) "characteristic of normal human thought," (b) positively related to the ability to care for others, (c) positively related to happiness and contentment, and (d) positively related to the ability to engage in productive, creative work. Conclusions a–d are important *if* true. Note that there are really three different claims: First, people characteristically hold beliefs that are exaggerated and unrealistic; Second, these exaggerated and unrealistic beliefs are normal; Third, these exaggerated and unrealistic beliefs produce good personal and moral effects (a–d). Let me lay out Taylor and Brown's experimental findings by looking at the evidence for claims 1–3; I'll then comment on these three claims just distinguished.

Here are some of the experimental findings for unrealistically positive views of self:

1. Given a list of trait names, subjects judge positive traits to be overwhelmingly more characteristic of self (and intimates) than negative traits.

2. Subjects rate self and self's performance on a task more positively than observers do.

3. Persons score themselves (and close friends and loved ones) better than others on all measures.

4. Persons judge the group or group(s) to which they belong as better than other groups.

5. Persons have more trouble recalling failures than successes.

6. Recollection of task performance is often exaggerated and remembered more positively than it was.

7. Favored abilities are seen as rare. Disabilities are seen as common.

8. Things persons do poorly are judged less important than things at which they are accomplished.

9. People think they have improved in abilities that are important to them even when their performance has remained unchanged.

10. Initially modest attributions of success or failure become more self-serving over time; for example, on a joint performance, credit given to partner gradually shifts to self.

Comments:

a. I am not handy. I am aware of this and judge it to be a sometimes inconvenient but not a very important feature of me. Is there a mistake? No, in part because there is no objective fact of the matter about how important it is for each individual to be handy.

b. Memory and narrative are interconnected, but they can be distinguished. I may remember various embarrassing failures well enough, but I would be a fool to dwell on them and even more of a fool to incorporate them into the self-revealing narratives I tell to others. I don't incorporate too many embarrassing tales into the stories I tell strangers or new acquaintances, but I do remember the embarrassing episodes. I want to look good and I want to protect my self-esteem. But these facts pertain to my desires, not necessarily to my beliefs. To the degree that narratives involve misrepresentation (which is different from accentuating the positive), we can be accused of lying or concealing the truth. But if there is something wrong here, it is a moral failure, not an epistemic one.[31]

c. The fact that I judge my performance as "better" than others do can be explained by three features of performance that don't involve false beliefs: first, as in (a) above, it is not clear that there is any objective fact of the matter about how well any individual does on many of life's tasks so long as she completes the task; second, I care more than others about how I am doing and thus I am paying more attention to how I am doing than others are and my attention may accentuate the intricacies of my performance to me, and this recognition may, to a point, be linked sensibly to pride in my performance; third, when I act I start with some first-person standards of success, I may adjust these some as I act and as I respond to particular features of the task. Others are often not positioned to see or notice these features, nor do they much care about noticing or attending to them. They are doing their own thing. Such facts account for some of the self-other asymmetries listed above without requiring the ascription of a powerful need to create positive illusions.

The findings related to illusions of control are as follows:

1. People often act as if they had control in situations that are determined by chance—as if, for example, skill at throwing dice mattered.

2. One's degree of control over heavily chance-determined events—for example, the sex of a child—is vastly overestimated.

Comment: The evidence for these illusions of control strikes me as especially dubious. Yes, everyone says "come on baby" when rolling dice. This is a fun little ritual that bespeaks hopefulness, not belief. Since I first came across these claims (more than 15 years ago) I have asked hundred of students about them; I have even bothered people at gaming tables in Costa Rica and Las Vegas. I have yet to meet *anyone* who really believes they have any control over such chance events, except in the baby case for those who know that there is some correlation between gender and vaginal pH levels, and that this can be controlled. I infer along with my students and fellow gamblers that if there are people who hold such illusions the correlation is with stupidity not with either mental health or happiness.

Finally, these are the findings on unrealistic optimism:

1. When asked their chances of experiencing a wide variety of negative events—for example, auto accidents, job trouble, illness, depression, or being the victim of a crime—most people believe they are less likely than their peers to experience such negative events. "Because not everyone's future can be rosier than their peers'," Taylor and Brown write (p. 197), "the extreme optimism individuals display appears to be illusory."

2. "Over a wide variety of tasks, subjects' predictions of what will occur correspond closely to what they would like to see happen, or to what is socially desirable, rather than to what is objectively likely.... Both children and adults over-estimate the degree to which they will do well on future tasks ... and they are more likely to provide such overestimates the more personally important the task is." (p. 197)

Comment: The fact that predictions correlate with hopes and social norms governing desirable outcomes allows the interpretation that it is best to understand such predictions exactly that way, that is, as expressive of hopes not as expressive of beliefs. The epistemic standards for hoping and believing are sufficiently different that no mistake needs to be imputed unless the hope is really wild. Personal reflection: I ride my Vespa around town. This is dangerous. I have four inches of metal in my left wrist from an accident several years ago. At the Duke Medical Center, people who ride motorcycles are nicknamed "motor donors." Does my riding my Vespa around town bespeak any illusion on my part? Well, I know it is dangerous (more so than walking or driving my car). One might say my actions belie saying I have properly absorbed the relevant knowledge about the danger I am in. This may be true. I suspect I have trouble imagining how bad things could go. But, even if this is so, it is not clear what false belief I have. There is a difference between not seeing something clearly and holding a false belief. I do see the point of considering this counterfactual: If you, Owen, were to have a really horrible outcome, wouldn't you wish you had made a choice not to own or drive the beloved Vespa? The answer is Yes. If I think enough about this I may change my behavior. But I still don't see that this would show that I have any false belief now. I might be accused legitimately of being foolhardy, inconsiderate of those who depend on me, and an imprudent risk-taker, but I don't see how any of this shows that I have any false beliefs. The point is being correctly ascribed foolish hopes and moral insensitivity are different failings than holding false beliefs.

In any case, consider each comment above as a caution against buying into any hyperbolic or global assessment of what the research on "positive illusions" reveals. There is no doubt that we hold many false beliefs and that many of them are about us. But nothing warrants the global conclusion that all persons at all times are destined to harbor false, self-serving, positive illusions about themselves.

Taylor and Brown ask whether any subpopulation can be distinguished from the general population merged in the meta-analysis, which is not prone to these findings:

> Does there exist a group of individuals that is accepting of both the good and bad aspects of themselves as many views of mental health maintain the normal person is? *Suggestive evidence indicates that individuals who are low in self-esteem, moderately depressed, or both are more balanced in self-perception. These individuals tend more* to (a) recall positive and negative self-relevant information with equal frequency ... (b) show greater evenhandedness in their attributions of responsibility for valenced outcomes ... (c) display greater congruence between self-evaluations and evaluations of others ... and (d) offer self-appraisals, which coincide more with those of objective observers.... In short, it appears to be not the well-adjusted individual, but the individual who experiences subjective distress who is more likely to process self-information in a relatively unbiased and balanced fashion. These findings are inconsistent with the notion that realistic and evenhanded perception of self are characteristic of mental health. (Taylor and Brown 1988, p. 196; emphasis added)

How ought we to understand these data? Consider the three claims I distinguished above: First, people characteristically hold beliefs that are exaggerated and unrealistic. Second, these exaggerated and unrealistic beliefs are normal. Third, these exaggerated and unrealistic beliefs produce good personal and moral effects. To which we can now add this: The only group not to suffer from exaggerated and unrealistic beliefs is the moderately depressed group.

The first question to ask pertains to the fourth finding, that the only group whose members are on average not prone to exaggerated and unrealistic beliefs about the self are the moderately depressed. What other groups besides the moderately depressed were picked out for study? Well, the very depressed were on the radar. In some other studies, the groups examined were extroverts and introverts. So we have at least five groups: normal, moderately depressed, very depressed, extroverts, and introverts. (Let's assume that the introverts and the extroverts are normal in the sense of "not depressed.") The first thing worth pointing out is that the very depressed, not surprisingly, do have exaggerated and unrealistic beliefs of a different sort that normals. They think that they are much more likely to suffer various calamities. If the odds are 1 in 6 that a male will get prostate cancer and I am very depressed, I might say that I will almost certainly get prostate cancer! Now in some studies in the meta-analysis such factors as education and SES are examined, as well as traits like neuroticism, and no statistically significant differences were found.

Let us analyze these four claims by adding to the mix some theory and data that have become better understood since Taylor and Brown published their paper.

- Infants as young as 10 months reveal a characteristic basal affective style as indicated by the ratio of LPFC:RFPC activity as well as response recovery to baseline from shifts caused by positive and negative stimuli. Affective style is indicative of mood and whether one is judged as an optimist or pessimist.

- Affective style has relations to certain personality traits such as extraversion and introversion, as well as consequences for such things as self-confidence, optimism v. pessimism. Affective style is neither illusory nor non-illusory; it is simply an indicator of mood and of one's cognitive-affective-conative orientation toward life. That said, optimists tend to think things, especially things that require their own effort, will turn out all right; pessimists are more wary and anxious about outcomes. Confidence is an important predictor of success.

- Positive psychologists refer to "positive offset" as the preferred and most common cognitive setting (again, one sees it in infants). When one starts the car one wants it humming evenly at a certain rpm so that take-off is smooth and reliable. Too low and too high are both undesirable and your car needs fixing. So too with persons. It is best to be ready to "get up and go." An experienced person knows what she is good at and she knows what she is not good at. I hire people to fix broken plumbing. Fixing plumbing is less important to my identity than certain other things, not because I judge it objectively less important than other things, but rather because, given that it is something I don't do well and others do, my doing it well doesn't figure strongly in my own identity assessments.

▪ There are no data that people with positive affective style (most people) are prone to holding false beliefs across all three of the domains.

▪ Studies show that the better-educated a person is the more likely they are to absorb and try to change certain, say, dietary habits, when they are given information about risk factors for various diseases. Thus thinking one has a lower than base-rate chance of suffering from a certain malady may have to do with the fact that one is thinking about or is in the process of making plans to adjust one's lifestyle in ways that one is taught will change the probabilities.

▪ Mind-reading skills come in two varieties: intrapersonal and interpersonal. Mother nature made people good at accurately reading characteristic facial expressions of conspecifics (Darwin 1871, 1872). And Mother Nature designed persons to be quick at picking up their own surface-structural beliefs and desires and executing actions accordingly. How good did Mother Nature make people at reading their own and other minds as required by contemporary social environments, and, what is different, as endorsed by Socratic norms? Well, one ought to ask first what people need to know about themselves and others to successfully negotiate current social reality. The answer is "Quite a lot." The answer varies depending on the complexity of the relevant worlds. But since we didn't evolve with traits for this modern social world, the best answer is that becoming a virtuoso at accurate self-knowledge, an adept mind-reader, and a deep knower of the nature of the external world requires the development of various knowledge acquisition strategies and skills that are not part of our innate endowment. For good Darwinian reasons, most people are accurate in assessing what is out there insofar as accurate perception and belief are required for negotiating the everyday physical world. To the degree that knowledge of facts requires more theorizing (water is H_2O, $f = ma$, thinking involves brain processes), 'knowing' defined as "justified-true-belief" becomes more difficult, more history and culture-bound, and thus less common. Such theoretical knowledge also generally matters less to fitness than knowledge of matters that simply require proper functioning of our basic mental equipment.

- Is the claim that people normally hold false beliefs, or is it best interpreted as that people generally have positive expectations and hopes—a positive attitude? 'Exaggerated' and 'unrealistic' are adjectives used to describe the whole set of epistemically questionable states of mind that show up in the meta-analysis. Every sensible person hopes she will not have various calamities befall her. But when such calamities, major or minor, occur no one but nutty people deny that the calamity happened to them. A sensible counterfactual test for whether a person in fact holds a belief in a strong and objectionable way would be: Does the state-in-question yield, and if so how quickly, easily, and so on, when there is countervailing evidence? Many, if not most of the alleged positive illusions will yield in the counterfactual situation, so perhaps they are not rightly understood as beliefs, at least not firm ones, in the first place. (I don't believe I will get prostate cancer. I get it and now I believe that I have it.) Second, there is conflation in Taylor and Brown's study between being an optimist or a pessimist and suffering from illusions. Optimism is clearly adaptive. It has an outlook component, I see the cup as half full, and an action component, when I act I plan things in such a way that I am more likely to succeed than fail. If I am playing a game and think to myself ("I think I can win") I do something sensible—I shore up my self-confidence—and by so doing increase my chances of winning (whatever these are). When I think this thought does it follow that I believe that [I will win]? No. 'Can' does not entail 'will'.

The main point is that even if the meta-analysis is taken at face value, two things must be kept firmly in mind: that no one holds all the illusory beliefs and that these are group effects. Thus, there are many individuals besides the alleged group of moderate depressives who don't live among the illusions. One task for eudaimonics is to find out what it is about such souls who reliably live outside the shadow of positive illusions that make them seek to live in the truth's full light, while at the same time assessing whether, and if so, what kind of trade-offs (if any) there are for such individuals regarding subjective happiness.

I close this chapter with several summary comments and observations and suggest a testable hypothesis that would, if true, show these results—or at least Taylor and Brown's interpretation of them—to be somewhat misleading.

- The effects, as I have said, are group effects, so we should want to look at the accurate self-appraisers who are in each of the five main groups (normals, moderately depressed, severely depressed, extroverts, introverts) who do not suffer from positive illusions and figure out why. Since there are some individuals in every group who do not suffer from the alleged illusions, positive illusions are *not* essential components of mental health or happiness. That is, if 'essential' is taken to mean 'necessary', it is just false that all mentally healthy people suffer from exaggerated and unrealistic views.
- It is obvious from a Darwinian perspective that ambulatory social animals need to have reliable urges—motivations to "get up and go." Such urges are best guided by figuring out the best strategies for success at whatever task is at hand and thinking that one will succeed. There is nothing illusory required for this.
- That said, Americans, especially, are subject to all sorts of social forces to spin their narratives (to themselves and for public consumption) in what are, on every view, excessively self-serving directions (see Frankfurt 2005) for a compelling essay on what the relevant phenomena require). This shows up in data that reveals considerably less self-serving self-promotion among, for example, the Japanese (Heine et al. 1999).[32]
- Other cultures don't encourage bullshitting and (what is different) self-deception as much as we do, so it is possible that we could bring the undesirable forces under control. One reason for optimism is that there is plenty of evidence that certain cognitive biases, such as representativeness and availability biases, recency effects, tendencies to think small samples are representative, affirming the consequence, and so on, that are ubiquitous across cultures can be overcome by training (Goldman 1986). The best explanation for these cognitive weaknesses is that Mother Nature did not design us as optimal epistemic creatures. But, as I have said, learning about these tendencies can noticeably weaken them. In the case of so-called positive illusions, they are not cross-culturally ubiquitous. This means that the degree to which they

exist in individualistic Western cultures is an artifact of their being socially encouraged. Because it is bad to believe what is false we should find ways to stop encouraging false belief. The objective standard for flourishing that I attributed to Aristotle in the last chapter would say this, as would twentieth-century existentialism with its emphasis on the virtues of authenticity. Eudaimonics can give some reasons for standing with these traditions and claiming that there is no necessary connection between positive illusions and happiness.

Here is the testable hypothesis: Study Americans who engage in Vipassana (insight meditation) on a regular basis. Vipassana has its roots in Theravadan Buddhist practices from Burma and Thailand, but has a totally secular form in America. It does preserve from Buddhism a determination to see things truthfully and honestly. Americans who engage in Vipassana are less likely to be depressed, but, more important, they will score much lower than normals on positive illusions. No psychologist has studied meditators with this hypothesis on the table. I challenge someone to do so. I will bet *all* the royalties for this book that you will find what I predict. The sample will have to be large and representative, the research peer reviewed and replicated. Then you will pay me, with all the money going to charity.

Appendix

Csikszentmihalyi (1997, pp. 111–113) lists the following characteristics of flow:

1. There are clear goals every step of the way.
2. There is immediate feedback to one's actions.
3. There is a balance between challenges and skills.
4. Action and awareness are merged.
5. Distractions are excluded from consciousness.
6. There is no worry of failure.
7. Self-consciousness disappears.
8. The sense of time becomes distorted (stops, speeds up or slows down).
9. The activity becomes autotelic, an end-in-itself.

Seligman says this about Flow:

On good days, "flow" is what happens to and for composers, musicians, athletes, dancers, rock-climbers. Flow is constitutive of "enjoyment," although enjoyable experiences are not high in emotion while we are in the zone.... Notice a salient absence: there is no positive emotion on the list of essential components. While positive emotions like pleasure, exhilaration, and ecstasy are occasionally mentioned, typically in retrospect, they are not usually felt. In fact, it is the absence of occurrent emotion, of most any kind of conscious awareness of one's state, that is at the heart of flow. Consciousness and emotion are there to correct your trajectory; when what you are doing is seamlessly perfect, you don't need them.... Csikszentmihalyi and his colleagues use the experience sampling method (ESM) to measure the frequency of flow. In ESM, participants are given pagers and then beeped at random times during the day and evening, and they record what they are doing at just that moment: what they are thinking, what emotions they are feeling, and how engaged they are. His research team has gathered more than a million data points involving thousands of people from many walks of life.... Flow is a frequent experience for some people, but this state visits many others only rarely if at all. In one of Mike's studies study, he tracked 250 high-flow and 250 low-flow teenagers. The low-flow teenagers are "mall" kids; they hang out at malls and they watch television a lot. The high-flow kids have hobbies, they engage in sports, and they spend a lot of time on homework. On every measure of psychological well-being (including self-esteem and engagement) save one, the high-flow teenagers did better. The exception is important: the high-flow kids think their low-flow peers are having more fun, and say they would rather be at the mall doing all those "fun" things or watching television. But while all the engagement they have is not perceived as enjoyable, it pays off later in life. The high-flow kids are the ones who make it to college, who have deeper social ties, and whose later lives are more successful. This all fits Mike's theory that flow is the state that builds psychological capital that can be drawn on in the years to come. (2002, pp. 116–117)

6 Spirituality Naturalized? "A Strong Cat without Claws"[1]

This is my last chance to say something intelligent with a worthwhile take-home message. My topic is spirituality and religion. Specifically, how can the scientific image of persons and spiritual and religious impulses, commitments, traditions, and institutions co-exist in the Space of Meaning$^{\text{Early 21st century}}$—if, that is, they can co-exist?

I will be using a distinction between asserting, on one side, and saying, stating, or expressing, on the other. Spiritual and religious conceptions cause themselves difficulty, as well as difficulty for those who are not on-board with a particular (or for than matter, any) spiritual or religious story, when their story is asserted to be true and authoritative. When a story (or stories) is stated, expressed, and understood as a story, those who spin the tale cause much less epistemic difficulty for themselves and others.[2]

First, a brief piece of autobiography. At the age of 7 or 8, I began to lapse from my Catholic faith: God just couldn't be as I learned he was. I was a skeptic, perhaps an atheist, if you can be one that young. I was not an agnostic, which even then I understood to be a sort of neutrality on the God question. The God I had learned about was not sufficiently loving or compassionate for me to think Him possible as God. I was in a bind, a painful bind, since "hell fire" did have me very scared. In 1966, in my first year of college, I read Alan Watts, D. K. Suzuki, and, best of all, E. A. Burtt's *Teachings of the Compassionate Buddha*. I was not then exploring becoming a Buddhist or anything else, but I was impressed by this: Buddhism was nontheistic and ethically very beautiful. I knew a fair amount about French deism, so I was familiar with the "Enlightenment" idea of conceiving of *my* God as creator and leaving it at that. I saw the merit in this, but I was not impressed by the ethical beauty of Enlightenment thinking. Politically it was good, but ethically, as a way of locating a noble, worthy, and meaningful way of living a life, I felt that something was missing. Enter Buddhism. The missing part was there. I was vividly aware that Buddhist practice, *if* there was in fact such a genuine item, truly embodied the only good idea I had ever heard from Catholicism—the Golden Rule, a rule that I was painfully aware was not the central part of the life of any Roman Catholic I had ever met, except possibly my mother, including even popes. God, remember, put kids in hot oil for all eternity if they died without having confessed their "impure" thoughts or actions. Jesus of Nazareth, rightfully the hero of the New Testament, truly believed in the Golden Rule. But my God was neither loving nor compassionate, not even to innocent children.

I was existentially cut off from my God at a very young age. I remained a good altar boy and even a compliant Catholic boy. The Mass in the late 1950s was still in Latin, which was good. We wore exotic, often colorful altar boy uniforms, and there was enough incense during certain services to keep me curious about this zone of life. I was quiet about my existential situation. I often served 6 A.M. daily Mass and helped give communion to the same old red-faced men and old non-red-faced women. I loved these old folk, thinking they were sweet, dear, and needy. I was cut off and worried. I had, without knowing it, a Nietzschean conviction toward the sweet old folks: "Your God isn't." But I felt protective of the old folk in a Dostoevskyean way: I would hold the secret for them. The Dostoevskyean thought that "if there is no God, then everything is allowed" crossed my mind with some regularity. But it, unlike the Nietzschean conviction, never gripped me. It seemed to me that there were plenty of reasons to aim for goodness independently of the mean-spirited system of reward and punishment that I was encouraged to believe in.

My father suspected that I was lapsing. He didn't know that I *had* lapsed. He was a very devout Catholic, the first in our family to go to college (Fordham University, after the Second World War, at night), and at the same time, an independent thinker in some domains. A copy of the liberal Jesuit magazine *America* was always on the table next to his chair, next to his martini. I would peruse it sometimes, realizing there were different views about the nature of *my* God, as well as on the Church's function, role, and position on a myriad of theological and social issues. I was glad for the Catholics that they allowed themselves some space for thought and free inquiry. All of Pierre Teilhard de Chardin's books on evolution were on the small set of living-room bookshelves, alongside the *Federalist Papers*. I skimmed them. When I was 13, my father gave me an abridged version of Aquinas's *Summa Theologica* that he had read in college circa 1948–1950. I read all five arguments for the existence of God carefully, saw the logical preposterousness of them all, and remember wondering vaguely if you could make a living examining and teaching about the tradition and the fallacies. This thought, as I recall, did not involve any intention to "trash talk" and expose what I perceived as preposterous. I saw how important the tradition was for the many "true believers." Meanwhile, I loved my Catholic education. At Archbishop Stepinac High School I learned Latin and Greek. Before each Greek class we said the "Hail Mary" in Greek, which was, of course, temporally inane. In junior and senior year on most Saturdays at 7 P.M., I would drive my girlfriend Barbara to confession in White Plains so she could tell a priest about our antics at the "Kensico Reservoir Lover's Lane" from the previous Saturday. This was a bit nutty, since we were on our way back to the Kensico dam to continue our sexual education, and Barbara certainly would not be able to take communion at Sunday Mass. But it all made its own crazy sense.

Kids from Stepinac, almost all Irish or Italian Catholics, who did not intend to become engineers, went to Fordham, Boston College, or Georgetown. In 1965, there was no doubt that I was, despite being an atheist, a Catholic—an Irish Catholic—and I would go to Fordham University. This was fine; if fact, it was good. I received a great education, and no religious types ever tried to win me around. Ed Reno, my first philosophy teacher, a very strong and tall A.B.D. (All But Dissertation) from Yale, said these words on the first day of freshman philosophy class: "Plato posits *the* Good." I was smitten immediately by philosophy.

By Thanksgiving we had read Plato's early dialogues as well as his *Republic* and Marx's *Economic and Philosophical Manuscripts of 1844*. I felt as if certain things were falling into place. I had a particularly exciting Thanksgiving with Grandma Gus and Great Aunt Mame in Taunton, Massachusetts, reading Herbert Marcuse's *One-Dimensional Man*. By Christmas of 1966, I was a neo-Marxist Platonist, or some such inane adolescent thing. And remember Buddhism had caught my eye, but there was no one to talk about or study Buddhism with. In any case, as was my style, I was not pushing my ideas on anyone else. Remember, I was very Dostoevskyean, even though it would be a few more years before I read any Dostoevsky. I felt different, a bit lonely, but less and less frightened with each passing day. I smoked marijuana most nights after studying, I was very involved in the anti-war movement, but I never took a serious hallucinogen (other than hashish). No principles stopped me. I just knew too many friends who had bad trips on acid. I think this was unfortunate, since I knew even then that many very interesting states of mind, some that fit the description of spiritual, mystical, and so on, were drug-induced. A grown man who works in consciousness studies should know first-personally "what it is like to trip." Someday, in the right conditions, I plan to trip. I remain very interested in this question: Why do some altered states of consciousness make those who have them want to make assertions about more than the experience itself, especially sincere confident assertions about what is really real as based on these states?

My applications to graduate school expressed my intention to address a simple question, one that only a budding philosopher could think was an outstanding question worth exploring: "What is a person?" So here I stand, the grown man who was that boy, wondering about what a person is and trying to understand what religion is, how it fits in, and why it matters to so many.

Back to the Present and on to the Future

For the rest of this chapter, my exploration of spirituality and religion, and how they work within the Space of Meaning$^{\text{Early 21st century}}$ proceeds in this personal vein, although the overtly autobiographical interludes will now recede from view. It is part of my story that the early history that I have just said a few things about matters in this way: The theological and moral questions I first faced as a Catholic boy are still my problems, but I work on them now in the way they appear as deep but secular metaphysical and moral problems about the nature of persons and the meaning of life. I don't think—although it is possible—that my views on spirituality and religion can be explained as a reaction formation to my religious training. In fact, I like to think that the depth of my exposure to Christianity and my work on metaphysics, epistemology, and ethics might equip me to say some illuminating things about how to negotiate this space of meaning. Know that I see the issues from where I am and have come from. I will not pretend, I hardly have time, to speak either objectively or about all that is relevant to the status of spirituality and religion.[3]

How can the scientific image of persons and spiritual and religious impulses, commitments, traditions, and institutions co-exist in the Space of Meaning$^{\text{Early 21st century}}$—if, that is, they can co-exist? Recall the basic picture of the human quest that I drew in the first few chapters: We humans are cognitive-affective-conative creatures who live as beings in time, with our feet on the ground, interacting in and with the natural, social, and built worlds. Living is a psycho-poetic performance, a drama that is our own, but that is made possible by our individual intersection and that of our fellow performers in a Space of Meaning. How we act, feel, move, speak, and think in the world depends in large measure on how we weave a tapestry of sense and meaning by participation in various subspaces within the spaces of meaning that constitute the Space of Meaning$^{\text{Early 21st century}}$. I picked out six spaces as the relevant ones for my exploration where the spaces are ideal types associated with aspirations, theories, traditions, and images: {art, science, technology, ethics, politics, and spirituality/religion}. I also argued in the first two chapters that I thought Plato got this right about human nature: We have an orientation, an urge, to live among, what he described as the forms of "the true," "the good", and "the beautiful." The neo-Darwinian picture of our nature allows, I claimed, a similar picture without actually positing such things as genuine Forms of these things. It is, I suggest, this platonic orientation that captures part of what we are speaking of when we talk of our "transcendent impulses." Minimally, we seek to transcend a narrow, local, occluded view of the world, of our compatriots and ourselves, and to locate a wider, broader, deeper way of making sense of things and finding meaning. Where? In spaces that are truthful, good, and beautiful.

The space of spirituality and religion is unusually interactive, as the spaces go, with art, music, poetry, and the visual arts—with the aesthetic dimensions of life. From the fourth century through the eighteenth century this was—in the West—due to an explicit alliance between politics and Christianity and thus to patronage for arts of the right sort. Sacred music—Bach's "Magnificat" and his "Mass in B Minor," Handel's "Messiah" and his "Israel in Egypt," Mozart's, Verdi's, and Fauré's requiems, as well great religious painting, all do something to my heart and mind that is special and different from what non-sacred music and art do. I do not dare try to explain what that is or how it is done, although I do think it has something to do with activation of what many people are talking about when they refer to "transcendent urges and impulses." I do not think these impulses are all due to cultural learning, but I also do not think they are non-natural, supernatural, or otherwise spooky.

In any case, the "transcendental" or "spiritual" zone is one I go to on purpose sometimes, to have what happens happen. Emotions of awe, solemnity, a sense of the holy, sacred, and precious are evoked. In order not to think that I am going over the top by what I have just said, let it be known that I also sometimes go to a favorite transcendental place that recreates a hashish "trip" I had in 1970. By fluttering my eyes and listening to certain types of music through my iPod I can revisit the colored rings of Saturn that were (and still are) amusement park slides that I went to during that pleasant hallucinatory episode thirty-odd years ago (for the especially curious this does not now require any intoxicant).

The space of spirituality and religion is one of the few spaces in the Space of Meaning$^{\text{Early 21st century}}$, indeed across the Spaces of Meaning for any time and place, which has prospects for giving comprehensive form to life-on-the-ground. What I mean is this: Most people want and need to intersect and interact with all or most of the spaces in the Goodman set of {art, science, technology, ethics, politics, spirituality} in order to make sense of things and find meaning. But it is only some few artists, scientists, engineers, inventors, and professional politicians who are likely to find life in the relevant zones of art, science, technology, or politics both *the* most meaningful space for them *and*, at the same time, to find that space to be fertile and spacious enough to find in that space most of the fulfillment they need. Finding one space especially important for meaning is not all that uncommon. But finding that same space enough for all one's meaning-in-life needs is very rare.

A meaningful life for some very few individuals might come primarily from passion for, and fulfillment in, painting, poetry, composing or performing music, doing science, creating new technologies, or being politically engaged. For some very few, this might truly be enough.

There is one caveat, a reminder from Aristotle: most everyone, no matter what space they are most absorbed in and by, will also typically want friends, companionship, family, and perhaps passionate love. In terms of the Space of Meaning$^{\text{Early 21st century}}$ (which may go back, in the West, at least to the Space of Meaning$^{\text{5th c. B.C.E.}}$) this desire to connect meaningfully with others is a tentacle that descends from the space of ethics, since ethics is powerfully opinionated about matters of love, friendship, family, and community in living a good life.[4] But it also descends from the tentacles of politics and spirituality since these spaces are also rich in wisdom about human connection. And once again they interpenetrate with each other and the other spaces of meaning.

In any case, with respect to spaces of art, music, science, and technology where some are plugged in to a high degree *and* find satisfaction from an almost single-minded passion, most of us are much, much less involved than the aficionados, or better than the "absorbates" (to coin a term). We are appreciative and opinionated perhaps about art, music, poetry, literature, science, and technology, but we do not seek to find the main source of fulfillment in any one of these spaces.

The space of spirituality and religion is designed, it seems, to function comprehensively for those to whom it is designed to appeal—in principle to everyone in the vicinity. Among secular Western liberals, ethics and politics (and law) taken together, do similar work. That is, they can, in principle, provide the bases for what Rawls calls a "comprehensive theory of the good." There is a discussion underway right now on the American left about recapturing for itself some leverage among the spiritually inclined which has been more or less ceded to the religious right. One move is this: claim that secular liberalism only means that religion is not to be invoked as a basis for legislation. Simple Jeffersonianism. But a secularist in this sense can be religious—most are—and she has every right when engaged in political discourse to remind her interlocutors about what Jesus said about the treatment of the poor and suffering. Invoking what Jesus said, and what his rationale for saying it was, is not to seek to ground legislation in religion, specifically in Christianity. Jesus was a wise Jewish ethical thinker, a saint for all times.[5] If you think him admirable, then listen to what he said and why he said it.

What sort of theism can co-exist with the scientific image? What I mean by theism is a set of propositions about the existence and nature of God or gods. I state my view of the situation up front in this way: Aspirations to locate our selves in the vicinity of what is true, good, and beautiful are noble and worthy. Given our platonic orientation this aspiration is also natural (note: not all our natural tendencies are noble and worthy). Naturalism, as I conceive it, is plenty broad enough to make room for robust conceptions of the sacred, the spiritual, the sublime, and of moral excellence. But theism of the sort that takes certain texts as authoritative, that asserts that certain facts that cannot possibly be known by humans to be true are uncontrovertibly true, is a problem. *Assertive theism*, but not what I will call *expressive theism*, is epistemically irresponsible and dangerous to boot. Let me explain.

There is one form of traditional theism that can perhaps co-exist with the scientific image. Science does not answer the question "Why is there something rather than nothing?" The God of deism, or even a panoply of creative gods, can *seem* to do so. Since no answer is any good epistemically speaking—just differentially satisfying—go with some such answer if it satisfies you and site some sort of conception of meaning and goodness upon the foundation provided.

In *Breaking the Spell* (2006), Dan Dennett asks of religious belief, and specifically of belief in the God (or Gods) of the Abrahamic traditions, *Cui bono?* (Who benefits?). That is, who are the beneficiaries of religious belief? Dennett distinguishes these different beliefs: (1) belief in God and (2) belief in belief in God. One can ask who are the beneficiaries and what are the benefits of (1) and/or (2)?

Do most people (Americans, say) believe in God, or do they believe in believing in God, or both? The difference would come to something like this: recall the discussion of "positive illusions" in the previous chapter. I might not really believe that I have lower than average odds of suffering various calamities, but I might nonetheless believe that it is good for me to believe that I will not suffer these calamities.

One can imagine this answer to the *cui bono* questions regarding (1) and (2): Many benefit if they enjoy positing a satisfying story about why there is something rather than nothing. If that story is theistically minimalist in the sense that it keeps itself in front of the world, and in addition supports a way of living that is noble, honest, and worthy, then it is a good thing, and there is no conflict between what science says and what religion says.

Theistic minimalism of the *expressivist* sort involves the following: Place your God, or panoply of gods, outside the natural world as the creative force that brought the world into being. Most think the only option here is the thin, boring, disinterested God of the French deists. But that is not so. If one ascribes to the originary creative force enough power to create the world, then one can also, indeed one might as well, ascribe to it great goodness. The demiurge that overflows with goodness to make the world in Plato's *Timaeus* is a nice picture of the sort that is permissible. But since expressivism is not committed to truth, not even to reasonableness, the story can be as wild and imaginatively rich as one pleases. Forces of good and evil, multiplicities of gods—a transfinite number of divinities in transfinite universes creating transfinite numbers of new worlds through worm holes between universes enacting the most fantastic battles between forces of good and evil. Whatever you wish that feels compelling, satisfying, rich and deep. We are only talking about stories.[6]

A powerful creative force that is exceedingly good, or again a multiplicity of good spirits, provides both a satisfying answer to the puzzling question of why is there something rather than nothing, and at the same time, a way of grounding an uplifting moral vision that can provide a coherent sense of how to think, be, and live so as to find meaning.

Such stories might be thought to function as positive illusions as discussed in the previous chapter. However, unlike all the other positive illusions discussed and endorsed as mostly harmless and apparently helpful to living well, such creation stories might be thought to be blessed with immunity from all empirical or theoretical feedback. Remember illusions, even positive ones (assuming there are any), yield some to the facts. A creation story of the sort I am thinking about—if it is held to be literally true—is a delusion, or akin to one.[7] Whether a delusion can be "positive," all things considered, is something I am skeptical about.[8] But I will not offer any argument for my skepticism, other than to say it has its roots in deep-seated convictions that assertions ought to be testable.[9]

Theistic assertions might be sorted into three kinds: (1) the relevant propositional attitude (in this case a belief), (2) its content, e.g. [that an all-good and all-loving God created the universe], and (3) the epistemology that is believed to warrant the belief in that proposition. Commonly, the epistemology adopted protects the belief with the relevant content with a view about the authority of revelation (as God's word), plus sometimes first-person authoritative experience. Philosophically, the epistemic standards about revelation are losers, but as a sociological matter they are often impossible to budge. It helps a lot that entire communities believe in the epistemic authority of the relevant texts as the word of God. An additional important factor is that the epistemic standards governing certain other spaces in the Space of Meaning$^{\text{Early 21st century}}$ (art, music, and literature, for example) do seem to allow untestable expression. Friends of assertive theism might try to find solace from the example of these spaces. But religious epistemology of the worrisome assertive sort will not find friendly standards in these places for making its own problematic assertions since art, music, and literature conform best to expressivist analyses.[10]

That said, the overall epistemic situation is worrisome as a matter of philosophical sociology. Epistemology is an area of expertise among philosophers. It is not a general area of expertise among those who work in the special sciences although their research methods normally incorporate such expertise. Ordinary well-educated folk know little about epistemology. (It is a common experience to teach the cosmological or design arguments to bright undergraduates and show that they are invalid; only to have some students say they still believe in God *because* of those very arguments.) Meanwhile theists do claim to work in epistemology, but it is, as I see things, always in a question-begging manner, offering "epistemic standards" designed to warrant theistic conclusions. This is epistemically disgraceful; but thanks to massive social support, theists, via many priests, ministers, imams, and rabbis, get away with the ruse. Among the Abrahamic religions, Judaism is the only tradition that is remotely honest about its being a historical phenomenon.

But there is an acceptable move here. I have already hinted at it, and I find myself inclined to recommend it: Think of a very complex, exciting creation myth *as a myth*. If a creation myth is conceived as a myth then it is not a criticism that it is resistant to all forms of naturalistic confirmation or disconfirmation. We can conceive of such a myth as either speaking about matters that science is unopinionated about *or* as speaking about matters that science is opinionated about, but speaking about those matters in a artful, expressivist manner. Expressivist theism says what it says in a manner that recognizes itself, and conveys forthrightly that it is simply telling a story where (a) one is "wanted" and can come in handy for certain purposes (giving the youth some understanding of the complex texture of things), and (b) might do some good for those who want a story that makes the universe *seem* purposeful and not absurd, not an inexplicable given (one might judge the urges behind (a) and (b) as themselves irrational, but I will not take up that issue here).

That said, naturalism recommends, indeed it requires, an epistemically reserved attitude toward such origin stories, as well as toward any eschatological stories about what happens after one dies. That is, although naturalism is unopinionated about what, if anything, caused the universe to exist, this does not mean that anything goes when it comes to spinning origin stories. Because they are untestable, such stories can be said, expressed, even embraced, but they cannot be asserted as worthy of true belief. They are not evaluable in terms of the "true" and the "false." But you can like your story so much that you treat it as true, even if it can't be evaluated as such. Or at least, something like this might be the best way to describe the self-understanding of the persons who tell a certain story that they conceive mythically: They do not quite believe their story to be true (they can't responsibly do so), but they believe that belief in their story is beneficial.

Cui bono? Those who feel the need for the story and find it compelling, uplifting, and satisfying. Who might that be? Nearly everyone, since the questions grip us, and some epistemically very wild and crazy answers are exciting, compelling, satisfying, or whatever. The situation is not dissimilar to the common phenomenon of children who don't really (any longer) believe in Santa Claus, but go a couple of Christmas seasons acting as if the story is true. They put out carrots for the reindeer and cookies and milk for Santa. This is fun and makes the whole event especially meaningful.

The issues are more complicated when it comes to beliefs about miracles and the afterlife, as opposed to creation stories and myths about the nature of the creative force(s). Miracles are alleged to be happenings in this world, so they are evaluable by epistemic standards governing assertions about such happenings. Life after death occurs in another world (in the Abrahamic religions, the "next world"). This might seem to give afterlife stories the same immunity as stories that creation stories have. But this is not quite right. Such stories involve the possible futures of embodied sentient beings, and thus physics, chemistry, biology, and neuroscience get a say in judging whether survival makes sense. Most will say it doesn't. In the Hindu and Buddhist traditions we have a different issue. Many lives involve recycling back into this world after each death, by way of either reincarnations (Hinduism) or rebirths (Buddhism). Here the situation is closer to the miracle case. Miracles, reincarnations, and rebirths are all alleged to be happenings in this world, and thus they are evaluable by epistemic standards governing assertions about such happenings. And they will be deemed to be ill-founded ideas by those standards.

Recall that naturalism involves a commitment to a certain picture of the world and its operations—a metaphysic, that is anti-supernaturalistic, and this is so because it considers certain epistemic approaches that warrant belief in supernatural posits—in this world, at least—to be discredited. It would be a mistake, however, to think that naturalism is derived solely from the scientific image. Science certainly played an important role in the ascendancy of naturalism. But so have the successes of certain methods of imagining and locating persons morally, spiritually, aesthetically, and politically that accept naturalism this far: They reject a common epistemic foe that involves grounding certain beliefs and practices on certain texts that are deemed to be the word of God.

Plato, Aristotle, Confucius, and Buddha all rejected the latter sort of epistemology in advance of its becoming widespread and familiar in the West, and they did so for very similar reasons to those that humanists who live in a post-scientific world do. It is because the epistemology that privileges certain texts as containing how God sees things and what God commands is not benign. It can have horrific ethical consequences and thus it does not deserve respect and tolerance (Harris 2004, 2006). This is something that affects not only how we interact with religious fundamentalists but also has implications for how we treat the religious moderate. The fundamentalist and moderate differ in the following way: both believe certain texts contain the authoritative word of the One, True God. They differ only on the matter of literal interpretation. Genesis is literally true or it is metaphorically true. In either case, the belief that God is the creator is unquestionable. God's word, so conceived, is an action-guiding trump—either as stated literally or as "properly interpreted." Many will say that the gospels command universal love of humankind. This is true. But the first and greatest commandment is not that. Jesus taught this: "Thou shalt love the Lord thy God with all thy heart, and with all thy soul, and with all thy mind. This is the first and greatest commandment. And the second is like unto it: Thou shalt love thy neighbor as thyself." (Matthew 22: 37–39) The belief in authoritative texts that contain God's word—a view largely idiosyncratic to the three Abrahamic religions, Judaism, Christianity, and Islam—is supported by an unsupportable, predictably dangerous, epistemology. Theism of the sort that takes certain texts as authoritatively true is a problem—epistemically problematic and socially and political dangerous.

What Shouldn't Be Asserted[11]

Naturalism, as we have seen, places restrictions on what can be asserted legitimately, with epistemic warrant.[12] Asserting is different from stating or saying, in that asserting is governed by epistemic standards of warranted assertability. Stating or saying is epistemically free range. Self-conscious "myth-making" is an elaborate form of saying without asserting. You can say anything you want, including all manner of false and foolish things. It's a first amendment right. The restrictions governing asserting come from hard won victories by practices that use naturalistic epistemic standards, which are then refined, modified, enhanced, and confirmed as good knowledge-yielding practices because they are successful. Here are some things that ought not to be asserted (nor proposed as reasonable for others to believe true):

- *You should not assert that any creation story you believe in is true, or even that it is made up of "warranted beliefs."* You are free, of course, to entertain the story as an appealing myth. Rationale: Assertions need to be falsifiable in principle, and such stories aren't.

- *It follows that it would be irrational to demand, let alone expect, others to believe the same story you do.* Rationale: Epidemiological models in psychological anthropology show that religious ideas spread like germs inside communities (Sperber 1996; Lawson and McCauley 1990; McCauley and Lawson 2002). This explains shared belief. And such explanations, if reflected upon, undermine the rationality of expecting others to share your beliefs. People seem to be remarkably immune to noticing the "location, location, location" mantra most familiar from real estate agents, as it applies to religious coin. The mantra explains why folk attach incredible epistemic weight and value to obviously non-shared religious and moral beliefs.

- *Although you can't assert that your creation myth is true, you can assert that belief in its truth benefits certain folk. (This is exactly the move the Grand Inquisitor in Dostoevsky's* The Brothers Karamazov *makes: He gives the ordinary folk what they need—"miracles, mystery, and bread," and keeps the dreadful truth to himself.* Rationale: This assertion that belief in a false belief benefits folk, is a sociological-psychological claim and testable as such. Of course, a lot needs to be said about who the beneficiaries are, what sorts of benefits accrue, and so on.[13]

- *Do not give "supernatural forces" genuine causal explanatory force when making assertions of the form "phi explains omega." That is, do not assert that "Allah created the universe" is true.* Rationale(s): (1) The best explanation for why humans almost universally posit "supernatural beings" and /or "forces" to explain the unexplained (possibly in the case of the "Why is there something rather than nothing?" question—which may be "eternally inexplicable") is that we possess, as a gift of Mother Nature, an HADD, "Hyperactive Agent Detection Device" (Barrett 2000). How so? And what is HADD? The basic picture is this (Atran 2002; Boyer 2001; Bloom 2004; Dennett 2006): *Homo sapiens* have prepotent epistemic settings that are easily activated and that enable us to catch on to the regularities that govern the ordinary causal world, as well as ones that suit us to negotiate social life. Social life involves taking the "intentional stance," attributing beliefs, desires, and so on to other sentient beings, learning to read minds, e.g., to detect motives and intentions before they are enacted, to detect cheating and cheaters, and so on. In creating adaptations, evolution commonly overshoots. That is, designs that work very well in the domain they are designed for—in this case the "agent-attribution-domain"—will sometimes see agency where there isn't any—where, for example, ordinary causation is operative. Thus the human tendency to anthropomorphize. This tendency is normally innocent enough and causes no harm. Before the germ theory of contagious disease existed, and before much of anything could be done about germs, believing that evil spirits caused illness was neither a bad surmise, nor harmful (so long as people were following basic rules of hygiene for other reasons). (2) Supernatural interventions conflict with naturalistic epistemology: If, as often happens, an aggressive form of brain cancer in a person who is thought to have a short time to live goes into a complete remission, we often speak of a miracle.[14] One sense of 'miracle' is innocent: Owing to some odd constellation of factors involving the person's immune system, the nature of the cancer, and so on, she defied all odds. But some set of causal events produced the miraculous effect. Other times, the concept of a miracle is invoked non-innocently and it is not allowed as an assertion, as for example when a cure is attributed to Mother

Teresa or Pope John Paul II from beyond the grave. You can say "a miracle occurred" you just can't responsibly assert it. The objectionable usage involves causal commerce between immaterial causal powers and the ordinary physical world. The big problem is that immaterial causal powers can be attributed with impunity unless we introduce a testability constraint (specifically, a falsifiability constraint) in which case they can't be asserted at all. Two subspecies of "the big problem" are these, both seemingly insuperable: First, if immaterial causal energy is imputed to do causal work in the universe this violates well-confirmed conservation principles.[15] That is, any immaterial cause that is invoked to explain a material happening violates the laws of nature, because it assumes, indeed it invokes, the introduction of physical energy from a zone that contains no physical energy. This violates the principle *ex nihilo nihil fit,* which theists commonly deploy (but unsuccessfully) in cosmological arguments for the existence of God (Flanagan 2002). Furthermore, to assert (vs. say) that such tricks occur is to assert something unfalsifiable. Invoking a principle that says no empirical data are relevant to the truth status of the claim causes this problem. Second, if there already exists a standard causal account that explains the unusual occurrence, then the immaterial power invoked to "really" explain it overdetermines the event in a way that makes it explanatorily superfluous; its only warrant is that it permits saying (but not asserting) that a miracle has occurred.

Meaning and Moral Glue

I don't think that the HADD ("Hyperactive Agent Detection Device") account remotely explains all that needs explaining about spirituality and religion. It does explain a fair amount about why humans across cultures, at times even when cross-fertilization was not much in play, engage in supernaturalistic personification. But religion typically does more, much more, than provide causal explanation of why the world exists, or why surprising things happen where naturalistic causal explanations are, or once were, lacking. Two common functions are these:

1. Providing stories that make sense of things and provide meaning.
2. Providing some sort of "superglue" to an ethical conception.

Philosophers, for understandable reasons, focus mostly on the zone of religious assertions that compete—or seem to compete—with naturalistic causal stories, with stories about cosmic origins or miraculous happenings, as well as stories about what happens after we die. This can lead to the view that most of the work of a spiritual tradition or a religious one is mainly to provide such causal or eschatological stories. This could be an illusion, specifically an instance of an availability and/or representative bias, where one overrates the role or frequency of what one is paying most attention to. Suppose I am in the market for a new, small four/five passenger car, and I set my price range under $25,000. I now see lots of cars in my price range that fit my specs. Indeed, I see more than there are.

I am going to assume that what I will just call *theology* for simplicity's sake—a triplet of theories about creation, miracles, and the afterlife—is what philosophers pay most attention to, but that theology is not in fact all or most of what religions say, assert, or do.[16] What else is there? What other functions are there? My answer is "Meaning and moral glue." Furthermore, these two are commonly connected.[17]

In chapter 2, I discussed some recent work in positive psychology that claimed evidence for certain common, ubiquitous, possibly universal virtues. There was one alleged virtue on the list of six, "transcendence," that I claimed is not a virtue. It isn't because, as I said at the time, it doesn't have the right structure to be a virtue. A virtue is a disposition to perceive, feel, think, judge, and act in the way the virtue calls for when the right sort of situation(s) arises. Nonetheless, the fact that something in the vicinity of "transcendence" is alleged to be universal deserves attention. Recall "transcendence" is introduced this way: "The transcendent according to Kant is that which is beyond human knowledge. We define it here in the broad sense as the connection to something higher—the belief that there is meaning or purpose larger than ourselves." (Peterson and Seligman 2004, p. 38)

We are given two characterizations, one cognitive and one affective: Transcendence is a belief that [there is meaning or purpose larger than ourselves] *or* it is the sense or attitude or feeling that [there is meaning or purpose larger than ourselves]." Perhaps it can be one or the other, or both.

From what Peterson and Seligman go on to say, and putting what Kant says to one side, "transcendence," it seems to me, is best conceived as a prepotent part of our basic cognitive-affective-conative constitution as human animals that is easily activated across environments. It can then appear as beliefs, as belief-like things, as beliefs or belief-like things with feeling and motivational bearing, and across all these dimensions as weak or powerful "I know not what" thoughts or feelings of connection, merger/merging, awe, and the like. Transcendence so conceived has almost completely to do with such things as the urges to make sense of things and to live meaningfully. Both these urges involve, indeed require, situating myself in the world in some sort of expansive way, in a manner by which I become at once smaller and less significant than I am inclined to think, and more connected to and part of that which is large and great. Making sense of things and living meaningfully involve some sort of fulfillment of the urge to make sense of who I am, of the outside world, my place in it, and my prospects for living meaningfully as a wee part of everything that there is. Many people who say that they are "spiritual" but not religious are saying that they are seeking to understand and develop a sense of connection to that which is greater than and more comprehensive than their self. In this manner meaning is sought, possibly found and embodied in one's life. The spiritual aspirations of such an individual do not, however, involve any theological beliefs. The individual might go to church, but not to worship God.

Therefore, I will treat "transcendence" this way, as an urge, a prepotent disposition or orientation, and not as a virtue. And I will, at least for argument's sake, take seriously the idea that, in either or both its cognitive and affective forms, something like it is universal in this sense: one sees forms of "transcendence" in most every human community. How individual persons interact and intersect with the communal forms of transcendence available to them is variable.

Despite the fact that "transcendence," as I understand it, is not a virtue, it might support virtue. What I have in mind is a simple thought. The perennial question: Why be moral? might receive a credible answer in the process of individuals and communities searching for answers to such questions as these: "What matters?" "What counts?" "How do I (or we) find meaning and purpose?" "How is my life connected to the whole?" That is, the question "Why be moral?" might—indeed I think it typically does—receive a credible answer while asking these latter, connected, questions. Why be moral? Because living morally is a condition for a meaningful life.

There is something normal and natural about having the thought that living morally is a reliable or necessary condition for living meaningfully. Why? One reason is biological. If a parent is nurturing, then at a very early stage, human infants release oxytocin when they hear mommy or daddy's voice. Oxytocin is a feel-good neurochemical, and it is implicated in later life in trusting behavior and generosity behavior (Taylor 2002). Some orphans miss the critical stage for oxytocin release and tragically don't ever recover to a point where even if placed in loving families they feel good, welcomed, and trusting. Nor do they behave generously except for purely instrumental reasons. Other reasons for coming to see the connection between morals and meaning are biological *and* cultural. Play and games are fun, but they require cooperation. Learning about the norms governing cooperation in play and games is a laboratory for gaining experience and knowledge about the role of normative governance in other zones of life. Life is better, more fun, less boring, and more meaningful when one learns to play games well, read "cooperatively," than if one doesn't. School days go better if everyone behaves, and so on.

Then there are discoveries of talents. Some talents are energized by passion or reveal passions. Chess, art, music, and sports sometimes, even often, involve both—talent discoveries and passion discovery and energization. To the degree that all these activities are either $n+1$ (chess), . . . , $n+10$ (soccer) games or activities, or require as conditions of learning more than just me, my desires, and my activity, they require cooperative norms. Nothing very "moral" in the highfalutin or rhapsodic sense in which people like to employ the term 'moral' is required. But normative learning, some of it moral, is taking place.

Peterson and Seligman, as I have said, say that "transcendence" is a virtue. This is false. What they then do for "transcendence" is identical to what they do for true virtues: describe what they call "strengths of character" that are utilized by and give form to "transcendence" conceived as a virtue. I want to briefly comment of these "strengths" that are alleged to support or enable "transcendence," on the charitable assumption that what they find in the vicinity of the "transcendent impulse," as I conceive it, might shed light on how it manifests itself and what, if anything, this does as far as illuminating the nature of spirituality and/or religion.

The strengths associated with transcendence are each alleged to "forge connections to the larger universe and provide meaning." This is the list of four (I leave out one strength: "humor")[18]:

appreciation of beauty and excellence [awe, wonder, elevation]: noticing and appreciating beauty, excellence, and/or skilled performance in various domains of life, from nature to art to mathematics to science to everyday experience

gratitude: being aware of and thankful for the good things that happen: taking time to express thanks

hope [optimism, future-mindedness, future orientation]: Expecting the best in the future and working to achieve it; believing that a good future is something that can be brought about

spirituality [religiousness, faith, purpose]: having coherent beliefs about the higher purpose and meaning of the universe; knowing where one fits within the larger scheme; having beliefs about the meaning of life that shape conduct and provide comfort.

In what follows, I will draw these four orientations or attitudinal clusters (now not conceived as "character strengths") into a discussion of meaning and moral glue.

The Meaning of Life

The worst question to ask is "What is *the* meaning of life?" There is no single meaning of life. Persons naturally engage in psycho-poetic performances in which and by which they seek to make sense of things and to live meaningfully. This "questing propensity" is part of our natural cognitive-affective-conative endowment as social animals, specifically as *Homo sapiens*. It is part of natural teleology.

Returning to principles evoked and explained in the earlier chapters we can say this: Rawls's (and Aristotle's) Aristotelian principle is one way of describing what sorts of desires give shape to the human quest: There are basic, almost ineliminable, desires for companionship as well as desires to develop our natural talents and abilities. We then adopt Nussbaum's and Sen's Capabilities Criterion as a normative principle in order to fix the following sort of problem with going just with Aristotelian principle (AP). AP expresses a desire that ought to be honored, all else being equal.

Specifically, every human ought to be given equal chance to develop her talents and interests so as to live in a fulfilling and meaningful way. The problem is that the development of talents and interests depends on social recognition of them, and then on the availability of modes of advancing them. Before many important inventions (the violin, basketball, ice skates), people with talents for these things were in no position to recognize that they had the talents. The practices that could or would eventually support them did not exist. In many cases, this is just a fact of life, not something we can complain about. But there is, or can be, something in this vicinity to complain about. People sometimes, often because of sexist or racist practices, deny themselves rights to talents that (if they were not subjects of these practices) they would see as good for them. "Mistaken preferences" and "false consciousness" are possible, and when they occur, attempting to correct them is permissible.

Some argue that it is analytic that if a person prefers *p*, she prefers *p*. The Capabilities Approach says that a person may prefer *p* but not really, all things considered, prefer *p*. The Capabilities Criterion derives from the capabilities approach and says this: Honor the Aristotelian Principle, but when and where there are legitimate reasons to think that certain people are not "seeing" their human potential accurately, work to adjust their vision so that they are better positioned to accurately judge what they are capable of, what they want, and so on. This is all, I acknowledge, complicated stuff. In many cases, we need to ask individuals, groups, and whole societies to see how certain longstanding socio-economic-political practices occlude their own clear vision of their situation.

The Capabilities Constraint allows normative criticism within and across traditions. It justifies itself by utilizing now familiar insights from social psychology. These range from work on the authoritarian personality, to work on identifying with the oppressor, to research on "the hedonic treadmill," to common facts such as self-recognition of mistaken preferences based on (self-acknowledged) overrating of short-term pleasures, and so on.

Instead of asking "What is *the* meaning of life?" we can ask such tame but very difficult questions as these:

- How shall I (or we) live?
- What ways of being and living produce fulfillment and meaning?
- What attitudes and beliefs about such matters as my place in the universe is it sensible to adopt?
- How can I understand my life's meaning, given that I am mortal?

- Given what I know about my talents, aspirations, hopes, and expectations, and given what I know about the existing network of social support, what sort of sensible plan can I make about how to live?

I will not spell out my own answers. I can't do so fully; they are unfolding—works in progress. We are each, with social support, supposed to find our own way to the answers. Spiritual and philosophical traditions often do the following work: They give us a head start in asking and answering these questions by being repositories of past "good answers." Aristotle, Confucius, and Buddhism, all in different ways, see productive and respectful social relations as time-tested ingredients of meaningful and fulfilling lives. Aristotle, in his threefold taxonomy of friendships into friendships of utility, pleasure, and true friendship, where you love the other as yourself, says something wise that still resonates among those engaged in the quest. Gratitude, on Peterson and Seligman's, list, comes in backward-looking and forward-looking forms. If I see myself correctly, I see that I am who and what I am, thanks to ancestors. Furthermore, I want my descendents to have reason to be grateful that I existed and made some positive contribution to who they are.

Then there is the matter of death. Many spiritual traditions moralize death by embedding a life story in what, in chapter 3, I called a karmic eschatology. There is a payoff system that kicks in after you die. Your fate is determined by the moral quality of your earthly life, and in a surprising number of cases, by whether you believed in the "true God." This last idea can be said, but it can't be asserted.

Here is the way I think about my own death, given that I think karmic eschatology(ies) makes no sense (I don't even get why anyone would find the idea of living forever, even blissfully, very appealing). I recently heard a wise Buddhist friend say that "death is the ultimate absurdity, you lose everything you care about." This, it seems to me, is not true. Furthermore, it is not a particularly Buddhist way (even for a secular Buddhist) to see things. Here is a better way: If you live well, then when you die *you* lose nothing you care about. Why? Because *you* are no longer there. *You* are just gone. That which is gone has nothing to lose. That which was once something, but is now nothing, cannot suffer any loss. But assuming the world and the people in it, including your loved ones remain, then your good karmic effects continue on. This is something to be proud of and happy about while alive. Your goodness, your presence, your worth are why the living feel your loss, and are sad, possibly very sad. But you are not sad, you neither suffer nor experience any loss because you are gone. Nothing absurd has occurred. True, dying could be miserable, but your own death is nothing to worry about.

Moral Glue

Here we seem to face a chicken-or-egg problem. Which comes first: a theology, or an ethical conception? Does one ground or give rise to the other, and if so what is the direction of origination? My best guess is that because all social groups are made up of individuals with selfish urges, as well as tendencies toward fellow-feeling, that problems of social coordination will need to be addressed from the get-go with norms that prohibit big ticket moral violations, such as murder, taking more than a fair share of resources, and the like. And thus that such norms will be invented first. Inductive reasoning will identify certain things, including certain human actions, as dangerous or life-threatening or not—even if not in those terms—very quickly. That is, long before the HADD device kicks in and we wonder about why there is something rather than nothing, some moral norms will be introduced, abided, and violations punished.

But given tendencies toward free-riding, simple selfishness, and sensible knavery, social groups will utilize whatever systems of sanction for violations that might prove useful. Gods can usefully come into play to serve multiple roles. Gods or a God can seem to explain puzzling origin questions and, in addition, anthropomorphic spirits who see what is going on, even when humans, their policing mechanisms, governments, and so on, don't, can produce a superb normative incentive system. "He knows when you are sleeping. He knows when you're awake. He knows when you've been bad or good, so be good for goodness' sake."

The basic idea is such a natural one that it is almost inevitable that the creative function of theism will be bound to its moral function if only for earthy norm enforcement purposes, and then possibly consciously by cynical (or realistic, depending on how you see human nature) political leaders. Binding the two functions together is too good of an idea, too effective a means of getting people to do what is hard, not to be utilized. Thus it is so utilized, again and again and again.

I say that the binding of a karmic eschatology to a moral vision is "almost inevitable" because, well, it is not absolutely inevitable or necessary. Some small ancient communities, usually where there is infrequent exiting of members or entry of new members, have non-karmic eschatologies where ancestors recycle but without the elaborate system of heavenly or hellish payoff in another place. Among Trobianders, although this is rare, there is not even ancestral recycling (although the deceased do travel to an island 10 miles away). People are just supposed to behave well, and they seem to do so—to a point (Obeyesekere 2002). Furthermore, China, both ancient and modern—except possibly in Tibet—seems to be unusual in talking about heaven (*tian*) and heaven's mandate (*tian-ming*) in an impersonal way and without either a non-karmic or karmic eschatology. This seems to have been the pattern since at least the time of Confucius (600 B.C.E.). So, China, despite being famous for ancestor worship, does not have a canonical theory of whether or how ancestral spirits recycle.

That said, even in places and times where the lights of doctrinal theism are on the dim side, there are normally rites, ritual practices, sacred places, and ceremonies where people come together to enact and confirm their commitment to a way of living and being that is their own. Harvey Whitehorse (2004) helpfully distinguishes *imagistic modes* of spiritual practices and *doctrinal modes* (although they lie along a continuum and each utilizes features of the other). Shamanistic practices typically include high emotional impact initiation rites without a mother lode of doctrine. Calvinism involves austere and complex doctrinal views with very tame, one might say, boring rituals.[19] The first (older) imagistic mode is easily passed on, enacted, and re-enacted in small communities; the doctrinal mode, especially once literacy is common, can spread very quickly across geo-political space by way of textual transmission and word of mouth.

The Latin word 'religio' means "to unite or bind together." One problem with certain religious forms of binding together is that they have in-group/out-group structure. Even the golden rule is sometimes (possibly often) interpreted in such a way that "love of neighbor" involves love for members of one's own community only (Wattles 1996).[20]

The *superglue hypothesis* is the claim that theism is an unusually strong way to bind a moral (and perhaps even a political and economic) conception into some sort of unity. A theological conception might provide both conceptual glue, providing a unified *rationale* for the components of the moral conception, as well as an *enforcement* mechanism, a scheme of reward and punishment for normative conformity. The superglue hypothesis says we might need superglue, but it doesn't say what it needs to be made of, although it claims that it is normally made of what are or seem to be supernatural ingredients. But this is a sociological claim about what is typical, not a philosophical claim about what is necessary.

The superglue hypothesis is testable and involves such ideas as these:

- Moral compliance is hard. Any practice that promises rewards and/or punishments for compliance or lack thereof might do important motivational work.
- Moral compliance is hard. Any practices that remind individuals that they share a form of life, are co-dependent on each other, and so on, might do important motivational work.
- The moral emotions which include the "reactive" attitudes of guilt, shame, regret, contrition, remorse, resentment, and envy, as well as such positive emotions as awe, love, empathy, and gratitude, can be sensibly and powerfully unified by practices that tie them to a unified conception of a normatively approved form of life.
- Besides being evolutionary adaptations in their "proto-moral" form, the moral emotions when activated have a self-referential nature. Self-reflection or self-awareness about one's own feeling, thinking, and behavior is constitutive of many of these emotions. Thus in their socially advanced forms they are often referred to as emotions of self-consciousness or self-assessment. And this element of self-awareness connects the moral emotions to issues of self-fulfillment, living meaningfully and virtuously. Insofar as a spiritual or ethical conception of a purpose beyond my own self-serving (selfish) ambitions is in play, I feel the motivational tug to participate in the normatively approved form of life (that does the job of realizing the higher purpose(s)).

Spirituality and Universal Love

Suppose we accept some version of the superglue hypothesis, two problems remain: First, there is the question of whether, supposing some sort of glue is useful for binding the ethical conception, it is necessary. Can the glue be supported by the ethical conception itself or does it need to come from something more than that conception, something supernatural? The answer is this: although it is common to use theistic supernatural glue, it is not necessary. Second, there is the problem of spiritual and moral chauvinism. The problem is familiar: many spiritual-religious-moral conceptions portray themselves in a way that is self-congratulatory. They assert that this is the (one, true) way. Proselytizing religions actively seek to convert wayward souls. Even non-proselytizing traditions often look with mixtures of condescension and bewilderment at those who do not follow their path. Both have been known to kill non-believers, at least partly for that reason (there is often the gold too). Sometimes there are good and legitimate reasons to disapprove of the way(s) certain people live. But sometimes it is simple chauvinism. The big problem I address in closing that relates to this chauvinism has to do with the surprising narrowness of the conception of ethical community—specifically the scope of moral consideration—that afflicts many otherwise noble spiritual/religious conceptions.[21] Given the ubiquity of chauvinism and given that something other than supernatural glue can bind an expansive moral conception, it would be better to have the glue supplied naturalistically if such glue can in fact, as I think it can, bind the moral conception without the chauvinistic downside, without distinguishing among insiders and outsiders, the chosen and the fallen —especially if in addition it can provide whatever additional sense of meaning that those who are "spiritual" but not religious claim to seek and sometimes find.

To get a sense of the problem that concerns me, consider Stephen Post's recent work on religion and spirituality and human flourishing. Post explores this "integrated hypothesis" as follows:

> First, human nature appears to include a powerful spiritual and religious inclination toward worshipful union with a presence in the universe that is higher than our own; second, a fully flourishing individual and society must make room for spirituality and religion (i.e., spirituality organized around group worship, rituals, symbols, and acculturation) in their salutary forms; third, these salutary forms can be described in terms of the degree to which they result in unselfish love of others, centering on our shared humanity, rather than on some small fragment of humanity. (2004, p. 1)

Post's integrated hypothesis is an hypothesis of human ecology, of eudaimonics. It is not simply a descriptive psycho-social-anthropological-historical hypothesis, but a normative psycho-social-anthropological-historical hypothesis, part of eudaimonistic scientia. I will simply accept that the first two parts of the integrated hypothesis are true. People are powerfully inclined to join with others to affirm stories that give voice to a higher purpose than their own individual desires. Such communities of worship, or what might be different, communities that express communal purpose to achieve meaning and high moral ends, can provide worthy forms of personal and communal identity, as well as powerful motivation to live well and in a way that is meaningful. And thus such impulses ought to be respected. (The last point expresses a norm based on the observation of what typically works to bind communities and recommends that it is wise to acknowledge and judge the norm worthy.) I focus on the third component of Post's integrated hypothesis: that "these salutary forms can be described in terms of the degree to which they result in unselfish love of others, centering on our shared humanity, rather than on some small fragment of humanity."

First pass, this third component can be read descriptively: these salutary forms can be described in terms of the degree to which they result in unselfish love of others. True, these "salutary forms" of communities of spiritual-moral commitment can be described in these terms. And then what? Clearly what Post wants to say, but which, as a social scientist, he may be shy about saying, is this: A spiritual-ethical vision that endorses a high degree of unselfish love is better than one that endorses unselfish love to a lesser degree, all else equal. There are three major world traditions that I know of that endorse a very high degree of unselfish love. Two are spiritual traditions, Buddhism and Jesusism,[22] although neither needs to be conceived theistically or, what is different, supernaturalistically, and one is overtly secular, Utilitarianism or Consequentialism. Utilitarianism or consequentialism (I will treat them as equivalent) says that we ought for fully natural reasons seek to promote the greatest amount of well-being for the greatest number of humans and other sentient beings.

If these three are considered the most worthy contenders because they seem to satisfy well the third component of Post's hypothesis, it might seem that the first two, Buddhism and Jesusism will fare better, all things considered, because they satisfy the first two components better than Consequentialism does, in virtue of being transcendent spiritual traditions that are typically linked with some form of theology. Maybe that's wrong. How so? One might claim that Consequentialism has been picked up by certain liberal spiritual/ethical traditions that sometimes incorporate Jesusism, and to the extent that they know about it, Buddhist ethics as well. I have in mind the Society of Friends (Quakers), Unitarian Universalists, and the Ethical Culture Society. Furthermore, Buddhism and Jesusism do not need to be conceived, and are not conceived by all their proponents theistically or again, what is different, supernaturalistically. Normally Buddhists and Jesusists do come together to affirm their tradition, but they sometimes do so in ways that do not involve supernatural reasons or justifications for the expansive form of love they espouse.

Buddhism, Jesusism, and Consequentialism: The Possibility of Universal Altruism

In Buddhism, an excellent person (a bodhisattva, for example) aims at embodying or realizing the four divine abodes: compassion, loving-kindness, sympathetic joy, and equanimity. Anyone can, and everyone should, take the bodhisattva's vows. Jesusism is defined, first and foremost, by the "Golden Rule," as well as by associated instructions on giving as much as possible—even at the cost of reducing oneself to poverty and homelessness—to those in need. He said to them "When I sent you out without a purse, bag, or sandals, did you lack anything?" They said "No, not a thing." (Luke 22: 35) Consequentialism tells us that we are obliged to promote the greatest amount of well-being for the greatest number of people (or sentient beings) in the long run.

Two questions immediately arise for all three traditions: (1) Why should I live in this expansive ethical manner? (2) How could a community, possibly at the limit every individual, be motivated to live this way? The two questions are related, although the first asks for reasons whereas the second asks, as it were, for methods to produce the proper motivation.

With regard to question 1, we might say that one reliable way to flourish, to achieve eudaimonia, is to live in this expansive ethical manner. With regard to question 2, we might try saying that humans have the relevant impulses to want to develop the relevant motivation. We might. But the settings are not, on any psychologically realistic view, strong enough without support—rational support, emotional support, methods of moral education and socialization and the like, to produce anything like expansive love. But perhaps with the right work to turn up all the initial settings, the desired effect can be produced. Maybe.

In Buddhism, certain contemplative practices are designed to do some of the motivational work.

Here I provide two examples of meditations designed to enhance ethical motivation that do not involve theistic or supernatural reasons for so doing. Both examples are provided by the Dalai Lama and are based on a type of Tibetan meditation (*Tong-len*), which is a form of widely practiced "give and take" meditation. The first is designed to enhance compassion; the second works first on selfishness, then on empathy and love.

> (1) In generating compassion, you start by recognizing that you do not want suffering and that you have a right to have happiness. This can be verified or validated by your own experience. You then recognize that other people, just like yourself, also do not want to suffer and that they have a right to happiness. So this becomes the basis for your beginning to generate compassion. So let us meditate on compassion. Begin by visualizing a person who is acutely suffering, someone who is in pain or is in a very unfortunate situation. For the first three minutes of the meditation, reflect on that individual's suffering in a more analytical way—think about their intense suffering and the unfortunate state of that person's existence. After thinking about that person's suffering for a few minutes, next, try to relate to that yourself, thinking, "that individual has the same capacity for experiencing pain, suffering, joy, happiness, and suffering that I do." Then try to allow your natural response to arise—a natural feeling of compassion toward that person. Try to arrive at a conclusion: thinking how strongly you wish for that person to be free from suffering. And resolve that you will help that person to be relieved from their suffering. Finally, place your mind single-pointedly on that kind of conclusion or resolution, and for the last few minutes of the meditation try simply to generate your mind in a compassionate or loving state (Dalai Lama and Cutler 1998, pp. 128–129).

> (2) To begin this exercise, first visualize on one side of you a group of people who are in desperate need of help, those who are in a unfortunate state of suffering, those living under conditions of poverty, hardship, and pain. Visualize this group of people on one side of you clearly in your mind. Then, on the other side, visualize yourself as the embodiment of a self-centered person, with a customary selfish attitude,

> indifferent to the well-being and needs of others. And then in between this suffering group of people and this self representative of you in the middle see yourself in the middle, as a neutral observer. Next, notice which side you are naturally inclined toward. Are you more inclined toward that single individual, the embodiment of selfishness? Or do your natural feelings of empathy reach out to the group of weaker people who are in need? If you look objectively, you will see that the well-being of a group or large number of individuals is more important than that of one single individual. After that, focus your attention on the needy and desperate people. Direct all your positive energy to them. Mentally give them your successes, your resources, your collection of virtues. And after you have done that, visualize taking upon yourself their suffering, their problems, and all their negativities. For example, you can visualize an innocent starving child from Somalia and feel how you would respond naturally toward the sight. In this instance, when you experience a deep feeling of empathy toward the suffering of that individual it isn't based on considerations like "He is my relative" or "She is my friend." You don't even know that person. But the fact that the other person is a human being and you yourself, are a human being allows your natural capacity for empathy to emerge and enable you to reach out. So you can visualize something like that and think "This child has no capacity of his or her own to be able to relieve himself or herself from his or her present state of difficulty or hardship." Then mentally take upon yourself all the suffering of poverty, starvation, and the feeling of deprivation, and mentally give your facilities, wealth, and success to this child. Thus, by practicing this kind of "giving and receiving" visualization you can train your mind (Dalai Lama and Cutler 1998, pp. 213, 214).

While writing this chapter, I spent a week in silent retreat doing this kind of meditation. Why? Because I am committed to improving myself. Many folk on such retreats take the bodhisattva's vows (nothing exotic is involved) by committing to following the path that such practices are designed to take one further on. Why should anyone who is not already committed to Buddhist ethics, or Jesusism, or Consequentialism want to engage in such practices, let alone commit oneself to them? They are, after all, designed to produce a certain way of feeling and thinking, a certain type of expansive, altruistic motivation. If you don't want to be "brainwashed," beware. I use the term 'brainwashed' deliberately. Many good friends and acquaintances are now studying how meditation re-creates the mind-brain. It does so by reconfiguring neural circuits, rebalancing the relations among our cognitive-emotional-conative settings.

Indeed, viewed from a naturalistic perspective, all practices designed to make persons more loving, altruistic, and compassionate—whether they work on social structures, or through rational argument, emotional appeals, or through charismatic persons—work, if they do work, by utilizing our neuroplasticity.[23] If one is a naturalist this must be so. If one is a supernaturalist one will say something else is going on. What that is, or what that might be, is not clear. Why? Because we are social animals, fully embodied and living in a natural world.

Is Universal Love Too Demanding an Ideal?

One might not consider this an interesting question. Ideals are just ideals. The real question is whether humans could in fact live in a way that embodies universal love, compassion, and altruism not whether we can embrace or harbor unrealizable ideals. It is true that there are two questions here, one about the ideal and one about our prospects for embodying it. Both need eventually to be addressed. One reason is that an ideal that is too demanding, in the sense that it is psychologically unrealizable, might be judged as useless since it can gain no motivational grip at all (Flanagan 1991b). But the relevant ideals in the three traditions under discussion do seem to be attractive as ideals. They are embraced by some after all, and furthermore, some few, perhaps, have embodied the relevant ideals. And what is actual(ized) is possible.

The charge of excessive demandingness is commonly made by philosophers against consequentialism (philosophers typically pay no attention to Buddhism or Jesusism),[24] especially against the version of act-consequentialism that requires that for each and every action opportunity I have, I should do, or try to do, what will maximize the greatest amount of good for the greatest number of people. The first objection to a theory of this form is that it requires a psychologically impossible amount of attention to "each and every action opportunity." What are all my action opportunities at this moment? How many different actions are action-opportunities that I have at this moment? If an ethical theory requires me to give determinate answers to these questions, even if I am only required to unconsciously compute what all my action-options are (before choosing the most "optimific"), it asks something impossible, perhaps something that makes no sense.

There are familiar ways around this serious objection. Most credible versions of consequentialism define the good as that which maximizes welfare impartially construed, but then go on to suggest a variety of psychologically possible ways that the best state of affairs can be brought about. For example, the good will be maximized if people proceed to love their loved ones, extend benevolence to their neighbors, show concern for their community, and care about the well-being of all. In this way, circles of concern will come to overlap, so that each is the beneficiary of an "expanding circle" of concern. The usual move is to suggest that impartial good will come from the spreading outward of partial concern.[25] It is not certain that this is true. But it is testable in principle, and it is a popular idea, captured most visibly in bumper stickers that read "Think globally, act locally."

Without resolving the question of whether there is an adequate way for consequentialism to keep its distinctive character (defining the good impartially), without also being judged to be too demanding, it will be useful for present purposes to try to understand exactly what feature or features of Buddhism, Jesusism, and Consequentialism are thought to cause the problem of excessive demandingness. The feature most often cited seems to have do with the requirement that we live in an excessively impartial way.

To get quickly to the point, distinguish among these different senses or types of impartiality: (1) the belief that everyone has equal right to flourish; (2) equal love for everyone; (3) impartiality in action.

The belief (1) that everyone has an equal right to flourish is not psychologically too demanding. Furthermore it is a sensible belief. Here, I repeat the Dalai Lama's rationale for (1), from meditation 1 above: "In generating compassion, you start by recognizing that you do not want suffering and that you have a right to have happiness. This can be verified or validated by your own experience. You then recognize that other people, just like yourself, also do not want to suffer and that they have a right to happiness." Simple empiricism.

As stated, this rationale is not logically demonstrative. If it appeals or makes intuitive sense, it does so by simply drawing on some sort of desire for consistency in thought.[26] But the two relevant thoughts, that I don't want to suffer and neither do others, do not entail anything about "rights" in the sense familiarly used in liberal democratic societies, where a "right" is something I can demand be met. Even if I acknowledge that others have a right to happiness, nothing follows logically about my responsibility to assist them in their quest. (Consider the standard libertarian view: everyone has a right to happiness, but I have absolutely no obligation to help others achieve happiness; that's up to them.)[27]

One might try this line of argument:

1. If there is something I desire for its own sake and recognize that everyone else wants the same thing, then I ought to believe that everyone has a right to that thing.
2. Whenever I recognize that I ought to believe something, I believe it.
3. I desire to flourish (not suffer, be happy).
4. I recognize that everyone else wants to flourish (not suffer, be happy).
5. I ought to believe that everyone has a right to flourish.
6. I believe that everyone has a right to flourish.

This argument is valid or can be made so. If it is, then everyone who believes the premises must believe that everyone has equal right to flourish.

The smart point of entry for a critic who wants to challenge the argument's soundness would be to challenge (1), the major premise. It states a norm in the form of a conditional, and thus the critic can legitimately point out that its logical form as a conditional wears on its sleeve the fact that the consequent does not deductively follow from the antecedent. Since this is true, the proponent of the argument will need to supply good reasons for asserting that (1) is sensible, wise, and true. I think this can be done. But (1) cannot be given a demonstrative, theorem-like warrant. It will have to be accepted, if it is, as an excellent idea with the weight of reasons behind it. In any case, the sense of impartiality that consists in the thought that not only I have a right to happiness, but rather that everyone does, is sensible and common enough. Affirming the existence of such a right is not demanding at all.

Next, consider (2) equal love for everyone. The question that arises immediately is "Who conceives of commitment to impartiality—now interpreted in accordance with (1)—as believing that everyone has a right (the same right) to flourish—as entailing equal love for all?"

Jesus and Buddha might be thought to be likely suspects. John Stuart Mill tells us that the main message of his essay "Utilitarianism" is summed up in Jesus' "Golden Rule." This is helpful since Mill is not, in arguing for consequentialism, promoting a doctrine that is particularly "lovey-dovey." That is, I don't need to love you in one familiar sense of the term to believe that I ought to promote your welfare. What then does "equal love" mean? There are questions about the meanings of both 'equal' and 'love'.

Mill's Jesus is not asking us to have warm and fuzzy feelings when he tells us to "Love one's neighbor, as oneself." What exactly did Jesus mean? Biblical scholars agree that Jesus best clarifies what the Golden Rule means when a hostile lawyer asks him "Who is my neighbor?" Jesus' answer comes in the form of the parable of the "Good Samaritan." The Jews and the Samaritans were bitter rivals and worshiped different Gods. The story is that a certain Jewish man is robbed, beaten, and left to die in a ditch. A Rabbi first, and then a Levite (a man of lower rank than a rabbi who assists the rabbi in preparation and oversight of Jewish religious services) both pass by, despite seeing the badly injured man. Then a journeying Samaritan comes along. "And when he saw him, he had compassion on him, nursed him, bound up his wounds, put him on his own donkey, and took him to an inn where he nursed him overnight. In the morning when it was time to leave he paid the Innkeeper and said, "Take care of him and whatever thou spendest more, when I come again, I will repay thee."

Jesus asks "Which now of these three, thinkest thou was neighbor unto him that fell among the thieves?" And he [the lawyer] said "He that showed mercy on him." Jesus then said "Go and do thou likewise." (Luke 10: 25–37)

Assuming, as everyone does, that the parable clarifies the meaning of the "Golden Rule," what does it help us see? It helps us see that hatred (*dosa*) is a vice, as in indifference, whereas sympathy, empathy, and compassion (*karuna*) are virtues. Clearly the Samaritan doesn't feel the sort of "love" toward the injured Jew that he does for his spouse, parents, brothers and sisters, children, fellow community members—toward whom he may feel different kinds of love. Whatever love he feels here is an active responsive sort of love, but it needn't be conceived as supporting certain desires to be in a continuing relationship with the man he lovingly cares for. No matter how he feels in his heart toward the Jew, his loving action is "impartial" in a familiar sense. It is a "love" born of compassion and mercy that would motivate him to help *anyone* suffering in the same way. The Samaritan doesn't know the injured man in any way that could make him feel any special way toward his character, toward the unique person he is.

We have insufficient evidence, but we might think that the Samaritan is someone whose heart is filled with compassion (*karuna*) and loving-kindness (*metta*). It needn't have been that way, since some weaker form of fellow-feeling could motivate a person to help anyone in such dire straits. A difference might reveal itself if the straits were less dire, or a major inconvenience would result from providing assistance.

In any case, the "love" required by Jesus in the "Golden Rule" is decidedly not personal love in any of the familiar forms; it is not romantic or sibling or parental or communal love. It is best described as compassion or loving-kindness toward any and everyone who suffers. As I understand Buddhism, it recommends the same virtues, the same kind of love. Same for consequentialism. This helps considerably with concerns about the psychological realism of all three conceptions or traditions. Why? Because if the demand to "love everyone" required loving in some familiar senses of 'love', then it would simply be unrealistic to think there was enough of that sort of love—those kinds of love—to go around.

What makes such love "equal," or "impartial," or "universal"? Well, as we have seen, it is not because one feels the exact same kind of love for the man in the ditch that one feels for one's children or spouse. What one does ideally feel toward both one's loved ones and the man in the ditch is impartial in the sense that one wishes happiness and no suffering for them all solely on the basis of their shared humanity. Both because of how one is positioned and because of the special (additional) love one feels for one's loved ones, one might, possibly through the work of meditation on such scenarios as the Dalai Lama suggests, take some of the deeper features of those personal love relationships and feel them into the wider world. This would be, I take it, a morally healthy thing to do. Is it superogatory, above and beyond the call of duty? Yes, if we conceive of our duty in the narrow way supported by certain Christian conceptions or their secular philosophical offspring as endorsed by Kant. But it is not supererogatory if we see the sense in what Buddha, and Jesus, and consequentialism ask us to consider.

Finally, consider (3), impartiality in action. A and B are both in equally dire straits. Both are drowning. I am equally well positioned to help A or B, neither of whom I know, but helping one means the other will die. It is obvious I must help one, but which one? The answer is that in this case it doesn't matter. Flip a coin if you wish. But save one.

But suppose A is my child and B is a stranger. I know what I would do. Critics of consequentialism sometimes say that a consequentialist in virtue of recommending impartiality in action should do the coin toss. The usual and plausible consequentialist reply is that the world will go better—the good, considered impartially, will be maximized—if people maintain, and abide the obligations intrinsic to their special, partialistic human relations.[28]

Such, happily rare, dilemmas aside, consequentialists, as well as Buddhists and Jesusists, will rightly press us about our chauvinistic tendencies. In a world in which 20 percent of the people suffer in absolute poverty as defined by the World Bank, am I really doing as much good as I can, as I should? Start just by thinking about America—although our love ought to eventually extend beyond the borders of our nation-state. "Imagine," sounding like John Lennon, that well-off Americans were raised to feel compassion (*karuna*) and loving-kindness (*metta*) in the way Buddhism recommends. There are 35 million working poor in America alone (none living in "absolute poverty" as defined by the World Bank) and perhaps an equal number of unemployed poor. (Aristotle taught that you can't lack basic necessities and have prospects for virtue and happiness; Buddhists agree, although they conceive of the basic necessities required as made up of a more meager basket of goods than Aristotle did.) Suppose that, in addition to being raised to be compassionate and loving-kind, we also believed in moderation and quelled our avaricious tastes to have whatever we want and can afford. If I am raised this way, the poor and the suffering are ordinarily on my mind, and I want to act so as to alleviate suffering and help such souls become happy. What to do? I could contribute (large sums of now-disposable) income to local charities. I could work for political reform, progressive taxes going from say 60 percent for people in my income bracket to 99 percent for Bill Gates. These are kinds of impartial actions, or better impartial strategies for the greater good. The people I am imagining, good Buddhists, Jesusists, or consequentialist persons, don't feel bad that they have less. They feel good that Fortuna's hand in determining the fate of our fellows is weakened and that the hands of those whose hearts feel love and compassion are strengthened.

The only sensible conception of impartial action I have ever heard defended runs along these lines. This is true despite the fact that we are sometimes asked to picture individuals looking at their savings ledger asking "Exactly what ought I to do at this moment to maximize the impartial good?" And with this picture in mind, the practical impossibility of doing anything helpful, let alone truly good, bears down on us.

I am aware that in this section I have not directly answered the question of whether an ethic of universal love is too demanding. I have, however, tried to do so indirectly. We ought to believe that everyone deserves to be free of suffering and to achieve some sort of happiness. Eudaimonics endorses this conclusion. It is both a matter of rational consistency and a matter of understanding what norms will lead to flourishing. Working on compassion and loving-kindness and loving our neighbor as ourselves make sense. Doing so, possibly uniquely, holds prospects for making us happier than all the money or stuff in the world can. Furthermore, it positions us—in virtue of our belief about what everyone in fact deserves, and our greatly amplified fellow-feeling—to want to actively work for the impartial good.

Being a virtuous Buddhist, Jesusist, or consequentialist is not psychologically impossible. There actually exist exemplars of each form of life, and what is actual is possible. That said, universal love will take lots of work and practice. But the three traditions being discussed reveal several promising strategies. Meditation, concentration and prayer (to rightly orient one's own heart/mind), rational arguments, and charismatic exemplars can all do considerable work. Furthermore, none of these traditions require supernaturalist glue to bind their expansive ethical conception.

Usually when an ethical conception is charged with being too demanding, the charge revolves around the criticism that the demands it makes are perceived to be psychologically or practically impossible. But none of the traditions under discussion advocates any states of mind for the virtuous that are impossible to achieve.

Each tradition presents a vision of virtue and happiness. We are not told simply that we are logically obligated or compelled to follow the way of expansive love and compassion. What we are told is that by so doing we amplify, in healthy and rational ways, our most noble natural tendencies.

If we want to find personal happiness and to make the world a better place, following the path of universal love and compassion (*damma* in Pali, *dharma* in Sanskrit) is wise and noble. Indeed, it is the only strategy that will work. Well that is not quite right. It is the best strategy among ones that can work to bind an expansive moral conception that is also naturalistic. Naturalism is the best philosophical view, based on the evidence, so this is a distinct advantage.

One final thought: An ethical vision of the sort recommended is "transcendent" in the sense discussed earlier, but which many incorrectly think can only be captured by a distinctively theistic conception. Happiness, flourishing, and meaning reside in the vicinity of embodying ideals of the human good that connect me to a goal beyond my own personal desires, and that are inclusive of all actual and future persons (possibly of all sentient beings) wherever they are or will be, as well as to the Earth and the larger cosmos that is our home. There can easily be houses of worship where committed people gather to affirm this way of being and living. That is the "good news." And that is "spirituality naturalized." Amen.

Notes

Chapter 1

1. "Nam et ipsa scientia potestas est": And thus knowledge itself is power (Sir Francis Bacon). The Greek 'episteme' is, like the Latin 'scientia', a word that normally means organized defensible knowledge including technical knowledge and the knowledge of craftspersons. But "eudaimonistic epistemics" sounds odd to my ear. Use it if you wish.

2. One might think that eudaimonics will support using any means to make people happy, for example, engendering false but consoling beliefs. But the seemingly innocuous word 'reliable' in this first pass formulation is intended to mean something strong: genuine and truthful. The "truthful" part is not simply a philosophical stipulation. It is (or may be) a finding of eudaimonics that, in general and over the long haul, "the true" is better than "the false" for flourishing.

3. The modern period of philosophy of mind starts with this dialectic: Descartes sets the stage by defending classical substance dualism (which is already largely assumed to be true). One of his correspondent students, Princess Elizabeth of Bohemia, immediately raises the objection that there is no explanation for how an immaterial mind (*res cogitans*) could cause a body (*res extensa*) to act, or how in the other direction the physical world could get information to such a mind. Descartes has a benevolent God playing this role: If something surely exists or occurs, such as mind-body interaction, then God, being good, wouldn't allow us to be mistaken about it, and thus God must somehow allow for the right kind of causal relations to operate. Leibniz and Malebranche try to help the dualist cause. But the solutions are worrisome. Both suggest that the best answer to Elizabeth is to give up on direct interaction between mind and body while maintaining dualism. According to Leibniz, for each individual, God sets two clocks, a mind clock and a body clock. They are endowed with "pre-established harmony" and stay in synchrony over time. So when I decide to go to the concert, my body goes to it. That's the way an omnipotent God set things up. Malebranche's God is more active. He intervenes on each occasion to keep mental and bodily events aligned. The first view is called "parallelism," the

second "occasionalism." These are clear cases of supernaturalism at work. Dualists still insist that immaterial mind-body interaction occurs and thus there must be an explanation that accounts for interaction between immaterial substances or properties and physical ones. The dualist thesis remains alive and well. The reason I say that dualism is not progressive is that no one has been able to say anything intelligible in response to Elizabeth's question to Descartes: How is this sort of interaction possible? The philosophical naturalist, equipped with the Darwinian insight that we are fully embodied beings and the neuro-psychological insight that brain keeps showing up as the most plausible site for "mind," has yielded a progressive research program that makes "interaction" between a mind/brain and the rest of the body and the world intelligible.

4. In my 1996a, 1996b, 2000a, 2002, 2003b, 2006a, and 2006b publications, I advance the idea of "ethics as human ecology." I think of eudaimonics as broader than ethics but as incorporating any and all insights gained from doing ethics as human ecology.

5. I use 'person' and 'human being' as synonyms throughout. According to a line of philosophical thinking that originates with John Locke, a person is a member of the species *Homo sapiens* who has memory adequate to track his life over time and who traverses space and time in a continuous way as long as he lives. For Locke, a comatose or severely amnesiac human = member of the species *Homo sapiens* would be the same human being but not the same person. I am a Lockean. But here I use 'person', 'human', and 'human being' as equivalent because I am talking about members of the species *Homo sapiens* who are self-aware and have adequate memory to track how they feel, think, what they do, and so on. Such creatures also ask meaning-of-life questions. When does such questioning begin? By the age of 5 or 6 years, many children are wondering about lots of things that we think of as philosophical questions: When did the world begin? Will it end? What happens when people die? Why am I me and not you? Such questions have metaphysical and existential depth.

6. Sellars is careful to point out that the philosopher engaged in what I am calling "the philosopher's vocation" must keep his eyes on the way he philosophizes: "Philosophy in an important sense has no subject-matter which stands to it as other subject-matters stand to other special disciplines.... What is characteristic of philosophy is not a special subject-matter, but the aim of knowing one's way around with respect to the subject-matters of all the special disciplines ... a philosopher could scarcely be said to have his eye on the whole in the relevant sense, unless he has reflected on the nature of philosophical thinking. It is this reflection on the place of philosophy itself, in the scheme of things which is the distinctive trait of the philosopher as contrasted with the reflective specialist; and in the absence of this critical reflection on the philosophical enterprise, one is at best but a potential philosopher." (1963, pp. 2–3) I should add that many analytic philosophers think that Sellars's description of the philosopher's vocation describes a fool's errand. What is certainly true is that it does not describe what most philosophers do or aim to do. And that is

fine, for many philosophical problems are too technical to work profitably on them while at the same time doing what Sellars suggests.

7. The twentieth century abounds with essays and books that examine the conflicts among what I call "spaces of meaning" by examining powerful pairs of images. Sellars's work is paradigmatic among philosophers. But C. P Snow's *The Two Cultures* (1959) is the locus classicus among the general intellectual public for examining how two major spaces of meaning hang together or fail to—in his case, the sciences and the humanities. My 2002 book *The Problem of the Soul* is in this genre.

8. When Max Weber coined the idea of the "disenchantment" of the world, he thought that "disenchantment," insofar as it was a problem, had many causes: capitalistic economic arrangements that set everyone's sights on accumulating wealth as the way to happiness, deep economic inequality, industrialization and horrific pollution of the environment. But *Naturwissenschaften*—forget about the human sciences—played a role in "disenchantment." The magic, mystery, and majesty of the traditional narratives of the human situation are displaced by stories whose main actors are impersonal causal forces. Atoms whirl, galaxies move, gravity does its thing, and so on. Bacteria and viruses cause illness, not evil spirits of the sort a shaman might expel. Weber like many others saw the prospect that there might in fact be such a thing as *Geisteswissenschaften,* human sciences that were poised to attempt to explain human behavior without *Geist* at all, that is by denying that human action had its sources in anything like an incorporeal mind or a soul. Such things would go the way of phlogiston.

9. I use 'platonic' and 'platonically' (lower case) to mean something that is in the spirit of Plato's philosophy. In the few cases where I use 'Platonic', I intend to speak of Plato's precise view. Plato's idea of a world in which things that are true, good, and beautiful are so because they "participate" in the "immaterial forms" of THE TRUE, THE GOOD, and THE BEAUTIFUL is not a view that has appeal for a naturalist. But the platonic idea that what is true, good, and beautiful is attractive to humans is an idea I take very seriously and try to make sense of in naturalistic terms.

10. For a still relevant and comical picture of the philosopher Socrates with his head in the clouds—among Plato's Forms—see Aristophanes' "The Clouds."

11. There has always been dynamic interaction, often outright competition, among the spaces that constitutes the Space of Meaning for a people. Plato, for example, used politics and ethics to put pressure on artists to stop producing art that he thought corrupted the youth. The Roman Catholic Church used religion to shut up Galileo. Mao used politics against religion. Picasso's *Guernica* represented a powerful genre of contemporary art with a moral and political agenda. Sixties music tried to rock the American moral and political world. Liberation theology in Latin America waged war with dictatorships and with capitalism. I could go on. The fact that our present situation calls for narrative expansion and adjustment due to developments across various axes within and between the spaces of meaning that constitute the

Space of meaning$^{\text{Early 21st century}}$ is nothing new. What might be novel about our situation is first that the rate at which we are asked to re-configure our stories has sped up. Thus the occasional urge to beg "slow down, you move too fast" makes sense. Second, science now wields much more weight on the shape of our lives and our self-locating narratives than at any previous time. One reason that pertains to mind science especially, is that it promises to explain with some measure of precision, more precisely than say evolutionary biology, history, anthropology, and sociology, the proximate causes of thought, feeling, and action. Furthermore, since about 1990 mind science has set out explicitly to explain the nature of consciousness. This last fact makes many "closet Cartesians" (those still gripped by mind-brain/body dualism) anxious because, well, consciousness is something that makes humans special, is inherently mysterious, and thus not scientifically explainable.

12. The right number for "worlds" is whatever number of persons there are (leaving aside other sentient beings). This is just one way of saying we are individuals. Obviously we need to organize this vast multiplicity into types to make sense of things. There is an important lesson here about the limits of science, and possibly about rigorous conceptual thought in general. Science like every form of abstract thought deals with types and kinds, only rarely does it engage in "thick description" of individual instances of things (tokens of types). One familiar objection to the human sciences and the scientific image of persons is that they fail to capture the richness of my life. The reply is that it is not their job (Flanagan 1992).

13. 'Philosophy' is not on the list, but it could be for people like myself as well as for many, probably most readers. I do think of myself and my life—and I interpret others—in terms of the concepts and categories that are familiar to me as a professional philosopher. I should comment here on the presumption in some quarters that Philosophy (with a capital P) has a unique role as a sort of Supreme Court for adjudication of disputes within and among the six spaces of meaning or of the modes of thought, genres, and individual disciplines that constitute each space. The presumption about the unique critical role of Philosophy is Kantian and can be seen as retained to some extant by Sellars. (See note 5 above concerning Sellars's view of the philosophy's place among the spaces and the disciplines.) The current situation as I see it is this: Philosophy is no longer viewed, inside the discipline or outside it, as capable of playing this unique critical and transcendental role. First, most every space of meaning, as well as whatever modes of thought, genres, or disciplines comprise it, have *internal norms* for how critique should proceed in that space. Second, when there is horizontal conflict among spaces—imagine scientists say there is global warming and a certain political regime denies it—the problem of which critical and epistemic norms, scientific ones or political ones, should govern the debate seems irresolvable unless we can find some meta-level norm that can be defended as the one that will resolve the dispute truthfully. One might think that in order to engage in such meta-level inquiry we need to give back to Philosophy its traditional,

unique, transcendental, and ahistorical critical role, one external to or above the internal norms that govern inquiry within each space. The rationale would be that the norms internal to each space or to the modes of thought, genres, or disciplines that constitute each space will promote their own internal norms of critique in predictably self-serving ways. But this is not the way things are now conceived. Critical theory as it originated in the Frankfurt School (T. Adorno, M. Horkheimer, J. Habermas, et al.), "Theory" in Literature (F. Jameson, T. Moi, et al.), Science and Technology Studies (B. Latour, B. Herrnstein-Smith, S. Jasanoff, et al.), as well as certain modes of philosophy (continental philosophy, feminism, critical race theory, gender studies) all use multifarious techniques to resolve debates about what meta-norms, if any, can be responsibly defended as worth entertaining to evaluate some actual debate in the real world. These non-transcendental = immanent forms of critique begin not by consulting "Pure Reason" (there is no such thing—never was) but by generating critique from within the terrain of actual practices. Sometimes meta-norms are invoked which are based on past and/or recent historical experience with irresponsible use of norms, e.g., scientists who claim certainty when the history of science reveals perpetual fallibility, or politicians who maintain oppressive practices for self-interested reasons. It is no longer credible to think that the latter sorts of vertical or horizontal critical inquiry is uniquely Philosophical in the honorific sense that it is based on "Pure Reason," is transcendental, completely objective, unbiased = unopinionated about zones where normative missteps are common, and thus completely neutral. That said, any and all immanent critique could rightly be called philosophical so long as it proceeds from deep love and respect for wisdom and the desire to uncover the sort wisdom that might lead to liberation, enlightenment, virtue, and flourishing for all persons. Naturalism entails that all inquiry and critique takes place inside the world, and naturalism is friendly to immanent but not transcendental critique. Transcendental critique requires a "view from nowhere." There is no such thing.

14. There are many Buddhist texts, for example, the various teachings of *lam rim*, a prominent set of guided meditations from Tibet—"the path to bliss"—that make much of the importance of learning to live with, and thus need and want, almost nothing. Most philosophers, west and east, think one needs basic necessities to have any chance of flourishing. But Aristotle (and for all I know the World Bank) may set the bottom too high. In the Buddhist case, the poor folk who can ascend the path of *lam rim* need to be provided with basic sustenance (often in the form of *dana*—gifts from others) and to be positioned to receive education, to be instructed on how to follow the path. But if there is peace and if there are good teachers, there is need for little beyond clean water, shelter, and simple sustenance.

15. Science also excels in the "what you are made of" relation, the constitutive relation. But depending on the question and context other "spaces" excel in the constitutive relation too. The real numbers are made up of an infinite set, indeed a larger infinite than the natural numbers. This concerto is made up of four movements.

This statue looks bronze but has lots of copper in it too. This poem is all in iambic pentameter, and so on.

16. Of the six spaces, two—spirituality (especially in its religious and/or theological forms) and science—typically are the most comprehensive. Spirituality typically includes an ethical vision, so this is one reason why it seems comprehensive. The situation of science is different. Most people, even well educated ones, don't actually know much science. What they do know is that science in our time (well, the situation goes back to the seventeenth century) seems to offer opinions on matters that some conceive of as settled traditionally by great sacred texts. Thus, some fundamentalists of various religions cast science as satan.

17. My old friend Stanley Fish gives a humorous example of the dependence even of good writers, professional biographers, on formats or master narratives available in publics spaces of meaning. In lamenting what he judges to be the sorry state of contemporary biographies Fish writes this in an op-ed piece in the *New York Times* (September 7, 1999): "[O]nce upon a time biographers didn't have to invent connections because they came ready made in the form of master narrative models. The two most durable were the *providential model*—everyone lives out the pattern of mistakes bequeathed to us by the original sin of Adam and Eve; and the *wheel of fortune model*—every life worth chronicling is an example of the principle of the general rule that what goes up must come down." One might say, speaking as a Kantian, that the providential model and the wheel of fortune model regulated the behavior of biographers. The subject's life is seen as lived according to the terms set by the providential or wheel of fortune model, and the biography written accordingly. Fish contrasts this with the contemporary "undisciplined" biography, which simply depicts a life as a string of unconnected contingent facts (which of course maybe it is).

18. To get a feel for the myriad horizontal and vertical tensions within the Space of Meaning consider just a few examples. Science in our time is seen as causing trouble for religious belief. But mutual revulsion between Muslim women in burkahs and Western women in jeans and tank tops involve conflicts between mores that have nothing to do with any scientific matter. Nor, although many think otherwise, are such matters discussed in the Qur'an or in the Bible. The Qur'an does say that women should cover their breasts, which both burkahs and tank-tops do. This is a conflict about sexuality modesty and is part of a larger conflict about cultural values. Similarly, many Americans (I include myself) appreciate rap and hip-hop as art—a truthful artful way of depicting aspects of life that ought to be seen and depicted. Picasso's *Guernica* is art that depicts the horror of war. Rap and hip-hop are art—melodic, poetic, clever, often fun and amusing—that depict ways of being and thinking that run from the unusual to the depressing. Some think that rap and hip-art are not art for reasons that are internal to, say, their conception of music. Others think rap and hip-hop, even if art, are ethically repulsive and socially dangerous. The first attitude is a conflict within aesthetics the second is one between aesthetics and

ethics. I think rap and hip-hop are often musically good as well as morally and politically valuable. Or take the debate about the use of stem cells. Recent scientific discoveries make the debate worth having. But the debate itself is a debate about values: moral, political, and (I think) aesthetic values. Then there are debates within science. For example, within contemporary mind science there are heated debates about the nature and causal efficacy of consciousness, about whether mental properties are sui generis or identical to brain states, about the relative roles of affect and cognition, about whether and, if so, what kind of "free will" humans have, and so on.

19. I emphasize that insofar as there are conflicts and tensions across the horizontal spaces comprising the Space of Meaning$^{\text{Early 21st century}}$, so too there are abundant conflicts and tensions within each space, vertical tensions. Many people will point to physics as the science that has its own house in order. But this is a fiction. Physicists can't be said to know, in the sense of possessing justified true belief, what nature is or consists in. As I write, theoretical physicists agree that two of the biggest feats in twentieth-century physics are the theory of quantum mechanics and the theory of relativity. The problem lies in trying to fit the two together, which is impossible in three dimensions. In fact, the two theories if embraced simultaneously as true, rather than as instrumentally useful—with small stuff and big stuff, respectively—are inconsistent. Thus one of the biggest goals in physics today is to unify the strong, weak, electromagnetic, and gravitational forces into one unified force, or what physicists call the "Grand Unified Theory." The point is that physics is not unified and there is no agreed upon answer to how the world works or even what it consists of or in at the most basic level. This fact may have important implications outside the philosophy of physics. Insofar as certain philosophical doctrines depend on physics for what they mean, for example, the imperialistic and unrestricted doctrine I call "global metaphysical naturalism" or more tame and restricted theses like physicalism or materialism about mind, they might mean less than defenders of these views think. I refer to my view as *neurophysicalism* as a way of avoiding the issue of what physics says the world is made up of at the most basic level (including whether there is such a level). One kind of physicalism about mind fusses primarily about the stuff that makes up mental experience; another kind fusses primarily about the causal interaction among the stuff. *Stuff physicalism* says that mind is made up of the stuff that physics talks about. On the (surprisingly controversial) assumption that this is physical stuff (is gravity physical? is electromagnetism physical? are elementary particles or strings that have zero mass physical?), all mental events are physical. *Causal physicalism* says that on the assumption that there are bi-directional causal relations among mental events and bodily events then whatever kind of stuff makes up mental and bodily events it has to be the kind of stuff that can have, and thus does have, the relevant type of causal relations.

20. There is also some interesting work on personality types that do not suffer severe bipolar disorders or deep depression but who have an "over-inclusive" cognitive style and who are more creative than average (Abraham et al. 2005).

21. It is worth noting that science is on the list of practices that can be studied by the human sciences. And it is so studied in programs in the sociology, history, and philosophy of science. In my experience, most mainstream scientists ignore or disapprove of this work—usually claiming that extreme relativists who wish to undermine their claims to objectivity advance it. Whatever. This reaction among scientists who are themselves studied scientifically is quite general. The very existence, even the possibility, of human scientific scrutiny of our nature and practices familiarly causes dis-ease, unease, and discomfort among those scrutinized. There are worries about reduction, elimination, explaining away, undermining the significance of cherished practices, and so on. When Max Weber introduced the idea that a certain "disenchantment" with life was the price paid for modernity, he had science in his sights as one significant causal contributor. Science like a mirror reveals things as they are and perhaps we'd prefer not to see what she can reveal so plainly. Vanity and all that.

22. Because images involve imagining and what one imagines depends on what one attends to or is impressed by, there are disagreements about what the scientific image says. Sellars says that the scientific image trades in unobservables and that the manifest does not do so. This sort of idea makes a surmise about the scientific image that captures a mid twentieth century obsession, but which in hindsight seems stipulative. The Greek atomist view of solids surely is part of the manifest image.

23. Francis Crick opens *The Astonishing Hypothesis* (1994) as follows: "[Y]our sense of personal identity and free will are in fact no more than the behavior of a vast assembly of nerve cells and their associated molecules .... You're nothing but a pack of neurons." One matter that needs exploring is who speaks with authority for science? One might think that Francis Crick gets to say what the scientific image is? The scientific image is, after all, an abstraction from the sciences. Which sciences? What are the principles governing the abstraction even if one is attending to the right sciences? To say "you're nothing but a pack of neurons" is attention grabbing but seems literally false even for fellow physicalists. The reason has to do with the logical force of the words 'nothing but'.

24. Some of my best friends—Joe Levine, Dave Chalmers, Colin McGinn—are mysterians. Some philosophers who have picked up the term use it to disparage "mysterians." I don't. The term is designed to describe a position. I have reasons for believing that adopting the regulative assumption of "subjective realism" and then doing the nitty-gritty work it recommends can close the explanatory gap. The mysterians are dubious. Colin McGinn for one is grateful for the moniker. That said, I did, when I proposed the term, distinguish "old mysterians" who are Cartesians. Both the new mysterians and I agree that that view is a non-starter. I should mention that there are philosophers, Thomas Nagel, for example, who I referred to as a "principled agnostic." The principled agnostic takes the standoff between the "new mysterians" and "constructive naturalists" (this is what I call my team) as reason for neutrality.

The Dalai Lama's position might be described as a form of "wait-and-see principled agnosticism."

25. Two points: First, the term 'neurophysicalism' (although from here on out I'll often just say 'physicalism') is designed to make it clear that we are talking about mental events being realized in the brain, and at the same time avoiding all controversy about what the world as described by physics will turn out to be. Second, strictly speaking the whole person is the experiencer. "*I* see the yellow tennis ball." My brain doesn't. It is harmless, however, to speak of areas of the brain that compute or represent color, shape, motion and so on. Likewise, as John Searle recently put it: "I digest my food." That's the correct attribution. As far as where (the location) the digestion occurs, that would be in my digestive system.

26. Of course, there would be no concept of 'water' or 'gold' and no theory about atoms or molecules to express these facts. But imagine that God would know them.

27. I have always defended the view that phenomenal consciousness includes qualia and propositional attitude states.

28. Many think that the field of consciousness studies will yield a hybrid neurophysicalism: Some mental states types, sensations and perceptions, for example, will be revealed to map neatly onto (relatively) similar and (relatively) homogeneous neurological sites. Ergo, type neurophysicalism for these states of mind. But something closer to token neurophysicalism will be needed for other mental state types. Take depression across a population of individuals who are judged as experiencing the qualitatively same kind of depression, call it *D*. Suppose, not unrealistically, that *D* supervenes on low serotonin levels in some individuals, or on low levels of dopamine in others, on the ratio of several different neurochemicals in others, or on an above normal rate of neural activity in right prefrontal cortex in others. Here the same experiential type *D* supervenes on, and thus maps onto, four entirely different neural substrates, each of which are sufficient to produce *D*, but no one of which is necessary for *D*.

29. Not all naturalists are neurophysicalists. Some think that mental events need to be described or individuated widely in terms of stimuli and responses and thus are not simply in the head. This view, sometimes called externalism, includes folk who work in embodied cognition, as well as some classical functionalists. What matters is that these folk are still naturalists. On a *wide individuation* approach to digestion, the digestive process is not just in the tummy (even if it happens mostly there). A digestive episode begins with a voluntary act of taking food from plate to mouth. The full individuation (= identification) of a digestive episode requires a description of the biochemistry of the food, what happens in the mouth, esophagus, stomach, intestines, etc., including a biochemical assay of what is eliminated as waste. The same story applies for some mind scientists to Pierre's experience of the taste of a fine wine. On a wide individuation approach the taste experience is not just in Pierre's brain, rather it includes a visual component (seeing the wine on the table), motor

activity (reaching for the glass), sensation at the periphery (lips, tongue, nostrils), as well as (possibly) certain behavioral effects.

30. There is another, related way to make the point in favor of subjective realism. This way of making the point turns on paying attention to indexicals, in particular to pronouns. "I" is an essential indexical from the point of view of the subjective realist because it essentially and uniquely captures, or at least, it essentially marks the first-person feels that I have been discussing. Description and explanation in normal mind science is in an objective *third-personal* or *impersonal* idiom: When an individual organism, O, sees a blue cube, there is binding of activity in the color and shape sectors of the brain. Generalizations of this sort do not capture first-personal feel. They assume that there is first-personal feel but they don't capture it. First-personal feel is only captured by the subject of experience. And this is why the first-personal pronoun is needed to explicitly mark experiences. I say "explicitly mark" because sentences like "Blue cube, here, now" also do the job, but only if it is assumed that the subject is, as it were, marking herself as the site of the blue-here-now experience.

31. One possibility, and it ought not to be ruled out automatically, is that phenomenology does not really support either physicalism or nonphysicalism about mind. Mental experience seemed to William James (a great phenomenologist but not a consistent friend of naturalism) to involve what Buddhists would call "very subtle" physical experience, as it did for Maurice Merleau-Ponty. Regarding "libertarian free will," the idea only gets really firmed up in the West with Descartes. In the East, in Chinese and Indic (Hindu, Jain, and Buddhist) philosophy nothing like it is posited. This much might lead one to think that the idea (possibly both ideas) become intuitively or phenomenological plausible only from inside a certain tradition. I like this idea, and I think there is something to be said for it, but I don't pursue it here.

32. Notice that our legal system diminishes the degree to which a person is culpable for a crime in accordance with Aristotle's formulation. If an individual is ignorant of the difference between right and wrong, or compelled to do what she does by insanity, a brain tumor, or a gun held to her head, she is not considered legally responsible for her act; at least she is less culpable that if any of these situations did not obtain. One other point: I am focused on agency here more than issues of responsibility. Many say that if the scientific image is true then our practices of reward and punishment and/or the justification for these practices will need to change significantly. Probably true, and all for the good.

Chapter 2

1. I continue to be charmed by Phillipa Foot's approach, old (1972) and new (2001). I think morality as well as other normative enterprises consists of hypothetical imperatives that are constrained by our nature as social mammals. When ends can be taken for granted or are universally shared we state certain imperatives categori-

cally. We also state imperatives categorically when the following situation obtains: We don't ultimately care if you share our view that murder is wrong, but be informed that if you dare to try to kill someone you will suffer very, very badly. Some philosophers read Foot's (2001) positing of species- specific norms as normal, natural, or basic as a departure from the earlier self who defends the view of "morality as a system of hypothetical imperatives" (1972). I think the views are perfectly consistent. Here is why: The desire to stay alive is normal, natural, and basic. If (as makes sense) you seek to live well, then you (ought to) sensibly believe in and abide norms that protect this normal, natural, basic, shared value. But commitment to this normal, natural, basic, universal value and to the norms that protect it is not categorical. Both the desire to stay alive and the norms that protect even one's own life can be trumped if unusual circumstances arise where things of greater value than one's own life are at stake. For example, giving up one's life for loved ones or to protect one's country from fascist intruders is considered perfectly understandable despite the fact that a normal, natural, basic, universal desire is overridden. The best explanation I claim is that such cases have this structure: Eudaimonia is what we seek. Living a life of some reasonable length is a necessary condition for achieving eudaimonia. The logic of eudaimonia, however, is such that some things matter more than staying alive if one wants (as one dies) to maintain one's self-respect and to honor one's highest values.

2. Legitimate evaluation requires, at a minimum, adequate observation of the other's life, adept mind-reading skills so as to reliably surmise motivation, and a defensible set of normative standards (ethical, aesthetic, political, and so on). One might worry that all this will gain us is something like intersubjective agreement that a life is good, meaningful, etc., not that it is objectively so. My response, and I won't dwell on it, is that intersubjective agreement about flourishing, though less than Plato sought, is in fact all we modern-day "platonists" can hope for. And it is enough because the bases for intersubjective agreement are empirical. That is, the evidence for claims that an individual is flourishing, is good, speaks truthfully, and so on will be public, defeasible, and not mysterious.

3. My own intuition is that it is somewhat easier to state which ways won't work than which ones will work. This might be because we are sometimes thinking about possibilities that haven't been tested in the latter cases.

4. Hume continues: "We see, even in the present necessitous condition of mankind, that, wherever any benefit is bestowed by nature in an unlimited abundance, we leave it always in common among the whole human race, and make no subdivisions of right and property. Water and air, though the most necessary of all objects, are not challenged as the property of individuals; nor can any man commit injustice by the most lavish use and enjoyment of these blessings. In fertile extensive countries, with few inhabitants, land is regarded on the same footing. And no topic is so much insisted on by those, who defend the liberty of the seas, as the unexhausted use of them in navigation. Were the advantages, procured by navigation, as inexhaustible,

these reasoners had never had any adversaries to refute; nor had any claims ever been advanced of a separate, exclusive dominion over the ocean."

5. We know that Darwin read Hume's *Enquiry Concerning Human Understanding*, but to my knowledge we do not know if he read the *Treatise* or the *Enquiry Concerning the Principles of Morals*.

6. A bodhisattva is a Buddhist saint who takes these vows: "Sentient beings are numberless; I vow to liberate them. Delusions are inexhaustible; I vow to transcend them. Dharma teachings are boundless; I vow to master them. The Buddha's enlightened way is unsurpassable; I vow to embody it." Bodhisattvas differ from Christian saints in that the vow is not made to God. Buddhism is non-theistic. The vow is to oneself. It involves a pledge to do what one ought to do because it is good and because by becoming a bodhisattva one is in the process of realizing one's highest potential (growing one's Buddha nature). Doing this will, if anything can, lead to flourishing, enlightenment, nirvana (Shantideva) (Lopez 1998; Flanagan 2006a).

7. The Golden Rule is quite common across the great ethical traditions. Wattles (1996) is very careful on the meaning of the rule across traditions. Of special importance is the question of scope—i.e., whether the rule applies to all persons or only to all the persons one interacts with.

8. What do people mean when they say they are spiritual but not religious? One analysis is this: Spirituality involves an acknowledgement of the transcendent impulse, which leads to a truthful recognition that one plays a small role in the greater scheme of things and of one's dependency on other people, Nature, and the Cosmos if one is to stay alive and have prospects for flourishing. Living in the space of ethics conceived in minimalist conventional terms is not enough for flourishing, nor is being mentally healthy in the minimalist sense of assessing things accurately, not being too out of sorts mood-wise, and so on. Acknowledging dependency and co-dependency on that which is greater than one's self motivates one to acknowledge in action the facts of dependency and co-dependency (MacIntyre 1999). A spiritual person denies that either moral or psychological goods conceived minimalistically, or even both taken together, are sufficient for flourishing. What else is needed? The spiritual but not religious person's transcendental impulse leads her, in the first instance, to acknowledge truthfully that she is part of and thus dependent on the whole—social, ancestral, and natural. Second, this truthful acknowledgement of her situation motivates her to live in a more attentive and expansive manner that respectfully embodies the recognition of this dependency. A spiritual person so conceived, who is not religious, makes no theistic claims, espouses no theistic beliefs. Often such spiritual people reject theistic claims as childish, unimaginative, and sometimes dangerous. A spiritual but non-religious person may, nonetheless, go to church to affirm her sense of community, her commitment to doing good, and so on (Goodenough 1998). I sometimes attend Unitarian-Universalist services for this purpose.

9. Assent to the Aristotelian principle will depend on how the clause emphasizing "complexity" is interpreted. Many traditions emphasize "simplicity" as a sensible way to lead a good life. There is no automatic conflict, so long as the Aristotelian principle is read as principled and restrained about the quest for complexity. One might think of the spirit of the bow to complexity as akin to Aristotle's view on true friendship. One can have only a few true friendships, otherwise one is spread too thin. Likewise, only a few of one's talents can be developed complexly. If one dedicates oneself wholeheartedly to developing one's musical, artistic, or philosophical talents, there will not be time or energy to develop all of one's other potential talents to a similar degree. What the claim then comes to is that once one starts to develop a talent, one will have urges to improve it, to take it as far as one can. Furthermore, Puritans, Quakers, and Zen Buddhists all agree that living the simple life is a complex undertaking, requiring difficult work on one's heart and mind, constant vigilance against the temptations for acquisitiveness, communal encouragement, meditation, prayer, and so on.

10. Once more I want to caution the reader not to read too much into the linear sequence of the narrative here. Aristotle does turn from ethics to politics, as does Plato before him in the later books (from III on) of the *Republic*. But both think that their own portrait of a virtuous agent can only be realized in a virtuous polity. So one could as easily and plausibly think that politics come first and creates the possibility conditions for good persons to evolve.

11. Capacities are emphasized over "functioning" because an individual might legitimately opt out of a capacity; e.g., a religious person might wish to fast at some detriment to his health. Nussbaum thinks that the language of capacities has an advantage over the language of rights insofar as it is not contentiously "Western."

12. Strict biological classification à la Linnaeus admit of a maximum of two Latin words. So we were dubbed *Homo sapiens*. Biological anthropologists and archeologists, not bound by Linnaeus's rules made discoveries which revealed for certain the temporal overlap of various *Homos*, several of which (e.g. Cro-Magnon) were very smart. As a result, several kinds were absorbed under the rubric *Homo sapiens*, and *Homo sapiens sapiens* is one of those.

Chapter 3

1. The first statement of this form goes back to 1981. (See Cabezon 1988.) Indeed, the Dalai Lama gives a remarkable example of what Buddhist beliefs might have to yield: "Buddhists believe in rebirth. But suppose through various investigative means, science comes one day to a definite conclusion that there is no rebirth. If this is definitely proven, then we must accept it and we will accept it." In a private audience with the Dalai Lama in Dharamsala in April 2000, he and I discussed the Roman Catholic conception of Heaven and Hell. He expressed concern about my reports of being scared by Hell as a boy: "Very bad." I then asked about Buddhist

views about the after life including karmic prospects for worse lives. He acknowledged that such an idea of rebirth was common among his people, but expressed personal reservations about his followers' idea of rebirth. I must admit to not probing very deeply—I was in a bit of shock. What is certain is that the Dalai Lama's conception of rebirth is more sophisticated than that of ordinary Buddhists. Buddhist scholars speak of rebirth rather than reincarnation, in part to express the fact that no person's *atman* can be reborn, or re-embodied. We are conscious beings but do not possess an *atman* = Self = Soul, an immutable, indestructible essence that can be reinserted in another body. Rebirth is the view that consciousness conceived along *anatman* lines goes on. How exactly is a bit unclear (to me).

2. The citation is to a book by the Dalai Lama (1990). One gets glimpses of it in other writings. But Jinpa's formulation is very clear and I will use it as the official statement of the caveat. He himself gives as an example the belief in rebirth. Science has produced no evidence for it, but has not shown its non-existence. A lot depends here on the logic required in "finding" or "showing." If the demand is deductive proof, then no rebirth is not ruled out. If it is a high level of inductive evidence disfavoring it, then arguably it has been ruled out.

3. At graduate school, in the 1970s, I was introduced to the problem of demarcation being played out in different ways among the logical positivists in the 1930s. Popper, from England, worked on the problem is his own critical way. Sometimes the problem was to demarcate in a principled way meaningful discourse that was scientific from meaningful discourse that was not scientific. So scientific assertions were testable, either confirmable or dis-confirmable, whereas the meaning of music, visual art, and literature was not testable, or better not testable in the same way, but nonetheless meaningful. Others were interested in a line of demarcation between sense and nonsense: science would fall on the side of sense because it made testable assertions, whereas (depending on the thinker) religion, Heidegger's philosophy, and wild and wooly traditional metaphysics would be designated nonsense. They seem to make assertions about the world, but they don't. Why? Because they are untestable gobbledygook.

4. It is a familiar logical fact that non-universal existential statements are not falsifiable in one strong sense. If I assert that unicorns exist, look, and fail to discover them, I can still, claiming a non-exhaustive search, stick by my assertion. Of course, I can be accused of irrationality if I have looked in all the right places and found nothing.

5. I would be foolish to deny that the doctrine of rebirth is considered mandatory across many Buddhist sects. My own expertise—insofar as I have one—lies in understanding Buddhism-from-a-Tibetan-scholastic-perspective. And again there are many who will say credibly that within that tradition it is mandatory. (See Willson 1987 for an excellent overview of "the logical arguments" the Dalai Lama accepts and advances in favor of the doctrine.) Perhaps, if it is not yet clear, I should state where

I am coming from: I think that Buddhism, broadly construed, contains a remarkably sophisticated metaphysics, epistemology, ethics, and philosophy of mind—the best, across the board, among classical philosophical traditions. But my overall naturalistic philosophical commitments lead me to want to tame what I consider implausible in Buddhism from that naturalistic perspective. This explains why I favor this statement by Stephen Batchelor (1997, p. 20): "I do not believe, as is sometimes claimed, that the teaching of Buddha stands or falls on the doctrine of rebirth, and that one cannot really be a Buddhist if one does not accept it." Robert Thurman's stated view ("Reincarnation: A Debate," *Tricycle*, summer 1997: 20–27, 109–116) is that to make the kind of spiritual progress the Buddha wanted peopleto make, you would have to be responsible for the sequence of former and future lives; but that you can be a Buddhist without believing this. I am not sure if Thurman intends to say that you don't need to believe (1) the "textual point" that this is what The Buddha thought; or (2) that you can question the view about the sort of responsibility one needs to take on to be sufficiently *motivated* to seek enlightenment and great virtue; or (3) that you can choose not to believe in rebirth, period. (2) the motivational hypothesis is very interesting and is faced and dealt with across most spiritual traditions.

6. My good friend Alan Wallace thinks that many devotees of these ideas have become ideologues. I hope he will find in me a non-ideologue who favors the views I favor because they seem most credible, all things considered. Of course, one should beware this locution "all things considered." It simply means "all things *I* (and possibly my friends, comrades) have considered/am/are considering." It is not even all that unusual to not be considering all one knows that might be relevant to the very problem one is considering. Maximal humility is always in order.

7. I have also had increasing contact with "Buddhism" in a very secular form through such people as Joseph Goldstein, Sharon Salzburg, and Jon Kabat-Zinn. Most of these teachers in "insight" meditation were trained in the older traditions of Theravadan Buddhism.

8. Rob Hogendoorn is now undertaking this important work.

9. See http://www.scienceformonks.org/.

10. A *geshe* is the holder of a degree granted for proficiency in debating. Becoming one typically takes at least 25 years.

11. See "A Huge Leap: Conversation with Achok Rinpoche," partly published in "De grote sprong: Gesprek met Achok Rinpoche" (*Vorm & Leegte*, winter 2005: 38–43).

12. There is a tradition among those who take the bodhisattva's vows—to live a life of complete dedication to compassionate and loving-kind action toward all sentient beings—to learn as much as possible about everything (including as many languages as possible), so all claims to knowledge can be tested with a critical eye. However, the Dalai Lama mentions warnings he has received that science and spirituality do not make good friends. He thinks, not implausibly, that this is a special problem for

theistic (Abrahamic) traditions. Rob Hogendoorn has helped me greatly to understand the position of the Dalai Lama among Gelugs as well as within the wider world of Buddhism. Hogendoorn explains: "Although the present Dalai Lama as spiritual and secular leader of the Tibetan people enjoys a considerable measure of respect and reverence, history shows that his standing even within in his own Gelug school does not preclude dissent. The monastic universities of Sera, Drepung, and Ganden traditionally conferred an unparalleled degree of authority on the *office* of the Dalai Lama; but their obedience to any particular *incumbent* is not unconditional." ("Westerse wetenschap voor Tibetaanse monniken," *Vorm & Leegte*, winter 2005, p. 34)

13. In Lopez 2005 I read that Chöpel called himself a *geshe* but never sat for the required exam. This fits his personality. Also in certain circles in exile Chöpel is a cultural hero or icon of Tibetan modernism (see ibid., p. 239). Perhaps I am safest saying that he is *persona non grata* among most monastic Gelugs.

14. Jinpa (2003) quotes from Gendun Chöpel's journal, written in the 1930s and the 1940s: "Even the Indian Brahmins who regard the literal truth of their scriptures dearer even than their own life, were eventually compelled to accept modern science." To move any recalcitrant Tibetan souls, Chöpel quotes from the revered epistemologist Dharmakirti (seventh century): "The nature of things cannot be cancelled, Through means of falsity, even if attempted, The mind will [eventually] uphold that [truth]."

15. In Bhutan, in Korea, and in Japan there are forms of Buddhism that do treat the historical Buddha or earlier Buddhas as a God. But this is decidedly not the norm across Buddhist sects throughout the ancient and modern worlds, where Buddhism is openly non-theistic. Many people, including Buddhists, keep this secret.

16. I am betting that if one were to poll ordinary Americans and Europeans about what impels scientific inquiry, most Westerners will say it is the search for truth, period.

17. Hopefulness is a good thing. But there are worrisome "laws" (really "generalizations") such as these: (1) whatever we can do, we will do; (b) whatever can go wrong (with technology) will go wrong.

18. Not all Buddhists are trained formally to debate. Gelugs are exceptional to the degree that formal debate, and practice in debating, is considered an essential part of their education and practice. Not even the other Tibetan sects practice debate to the extent that Gelugs do. That is, although there is some room for formal debate within other sects, it pales compared to that of the Gelugs. For Gelugs debate is integral to scholarship, for others it isn't. All Buddhist sects endorse, in principle, the examination of assumptions. In practice, much depends on how open a sect or a school is to critical examination of it. Of course, the same point about the variability of "openness to criticism" applies to science as well.

19. *New York Times*, September 18, 2005.

20. The oldest Vedas date to 1500 B.C.E. and do not include the *Upanishads* and the *Bhagavad Gita*. The *Upanishads* date from the sixth century B.C.E. The *Bhagavad Gita* is included in the *Mahabharta Epic*, written from the fourth century B.C.E. to the fourth century C.E. but the *Bhagavad Gita* is thought by many, perhaps most scholars to be a late text composed possibly entirely in the Common Era. In any case, the latter are the key texts of what came to be known as Hinduism. Hindus don't typically call their religion 'Hinduism' (although they may call themselves Hindus as a sort of ethnic attribution) the name originates most likely in the desire of British colonialist to name the religion/spiritual practices of Indians something! To make matters worse, the English word 'Hindu' is almost certainly based on a mispronunciation that relates to the importance of the Indus (not Hindus!) river. To describe their spiritual practice, Hindus sometimes use the word 'darshana', which is best translated as 'philosophy'. Often they refer to their way as 'Santana Dharma' ("the eternal way of truth"). There is no Hindu pope. It is not a creedal faith with a single orthodox doctrine. There is no Buddhist pope either. Buddhism is also not a creedal faith with a single orthodox doctrine. That said, every spiritual tradition has some commitments that constitute the minimal conditions of being a member, advocate, and so on. A traditional Tibetan textbook says: "The definition of a proponent of Buddhist tenets is: a person who asserts that the four seals which are the views testifying that a doctrine is Buddha's. The four seals are: 1. All compounded phenomena are impermanent; 2. All contaminated things are miserable; 3. All phenomena are selfless; and 4. Nirvana is peace." (*Cutting through Appearances*, p. 176) I'm not sure if other Buddhist sects—outside Tibet—define themselves using "the four seals."

21. *Brahman* is the name for the ultimate, self-sustaining source of all creation. But "it" is not a person. Furthermore, many Hindus conceive their elaborate pantheon of gods, even high Gods like Brahma (creator of Earth but not everything; that is Brahman's role), Vishnu (loving protector) and Shiva (fierce protector) as "aspects" on the one and only God, Brahman.

22. This led Bertrand Russell to claim that the cosmological argument specifically was self-contradictory; it rejects an infinite regress in the second premise, only to accept (a concealed) one in its conclusion. See Flanagan (2002).

23. The picture of a universe that is eternal, that has no beginning (is infinitely regressive) and no end (is infinitely progressive), seems to generate a Zeno-like paradox: How, in a universe that has no beginning and that goes back an infinite amount of time, did we get to now? I'll let the reader puzzle over this.

24. The Dalai Lama writes: "Even with all these profound scientific theories of the origin of the universe, I am left with questions, serious ones: What existed before the big bang? Where did the big bang come from? What caused it? Why has our planet evolved to support life? What is the relationship between the cosmos and the beings that have evolved within it? Scientists may dismiss these questions as

nonsensical, or they may acknowledge their importance but deny that they belong to the domain of scientific inquiry. However, both these approaches will have the consequence of acknowledging definite limits to our scientific knowledge of the origin of our cosmos. I am not subject to the professional or ideological constraints of a radically materialistic worldview.... And in Buddhism the universe is seen as infinite and beginningless, so I am quite happy to venture beyond the big bang and speculate about possible states of affairs before it. (2005, pp. 92, 93)

25. Many commentators say that, in the time of Buddha, Vedic belief involved the puffed-up idea that ATMAN was divine, the microcosmic twin of the macrocosmic DIVINE BRAHMAN. Spiritually (financially) advanced members of the Brahmin class were on their way to becoming liberated by being absorbed into that which they mimicked or mirrored.

26. When Descartes vows to doubt everything, he makes it clear that he does not throw his ethical convictions into the mix. As he sees it the "epistemological crisis" he is in does not require it.

27. Gratitude: In the West there is not—at least—any longer enough recognition of our epistemic dependence and our characterological dependence on our ancestors. What the Dalai Lama calls karmic causation together with the doctrine of (co-)dependent origination help one see reasons to have and express gratitude. One sees the right attitude in most Buddhist sects as well as in the Confucian tradition. Personally, I recommend ancestor worship.

28. I say "most" not "all" because if mathematics is non-empirical *a priori* knowledge then it might not for someone of Descartes' talent been learned from the senses or via authority. This helps explain why Descartes introduces the possibility of an "Evil Demon" to allow him to doubt the truth of mathematics in the *Discourse* and in *Meditations* I and II.

29. This quotation—"By contrast with religion, one significant characteristic of science is the absence of an appeal to scriptural authority as a source of validating truth claims" (2005, p. 25)—might be understood to be true if what we mean by validation is both validation and re-validation. But given the Buddhist claims to the priority of experience and reason over scripture it should be true in Buddhism too. We go to canonical texts, scriptures, classic scientific papers, textbooks, to discover knowledge that is considered to have been validated. All my students believe that water is H_2O. Testing this requires a non-trivial, and somewhat dangerous electrolysis procedure. A few of my science students have done so; but not to validate or re-validate that water is H_2O (although that too is being done), but to learn the experimental technique of electrolysis.

30. I have asked John McDowell to name names who hold the view he calls "bald naturalism." He has successfully evaded my request. I suspect McDowell has in mind the Churchlands when they were "eliminative materialists," although I don't

think that even in that stage they fit his description. But that's just an educated guess. As I read him, Alan Wallace (2000) interprets scientific materialism as scientific materialism^{imperialistic}. In the book and in person Wallace claims that the debilitating depression William James's experienced in his late 20s was due to his medical training, specifically to his introduction to the idea that mental events are/might be brain events. But the reason this got James into a very deep funk was due to the fact that as he understood the thesis it entailed epiphenomenalism (the quote from James on p. 159 of Wallace 2000 makes this clear). Mental events exist; but they do no work! This is a very disturbing idea, but it does not follow from commitment to naturalism about mind. On a view such as mine, a.k.a. "subjective realism" (see Flanagan 2002 and chapter 1 above) qualitative mental states are brain states *and* they are causally efficacious. I should mention that there is another, not inconsistent, theory about the etiology of James's depression. William—like the whole James family for several generations—was obsessed with the metaphysical problem of free-will and determinism. Libertarian free will and life's meaning were inextricably linked in his mind. (On how to decouple the two, see Flanagan 2002 and chapter 1 above.) See James's "Epilogue" to *Psychology: The Briefer Course* (1892) and the essays in *The Will to Believe*, especially "The Will to Believe" and "The Dilemma of Determinism." The combination of materialism and determinism as he understood both doctrines, supported epiphenomenalism *and* the denial of free will, ergo nihilism. Psychology, for its limited purposes, can assume both, but ethics (the more important discipline) must, for its purposes, deny both materialism and determinism.

31. For more elaborate criticisms of scientism, see chapter 1 above and Flanagan 2002.

32. Even if the Dalai Lama is not directly opening up Buddhist ethics to critique from science, he is in fact doing so indirectly. The reason is this: If science requires beliefs in rebirth or karmic laws to be rejected or modified, this has an implication minimally on the motivational structure of Buddhist ethics. I don't, however, see any issues in play that would affect the content of Buddhist ethics—that living in a compassionate and loving-kind manner will lead to flourishing.

33. I will assume that something like Ernst Mayr's *What Evolution Is* (2001) is understood by the Dalai Lama and his colleagues as an accurate representation of neo-Darwinian theory circa 2000.

34. Right now people in Bangladesh are vastly more fit from the point of view of population genetics than the people of Belgium. Why? Because there is a population explosion in Bangladesh and most of South Asia and zero population growth in Belgium. If all humans cared about was fitness we would all be trying to maximize the number of our own children, grandchildren, and so on. But flourishing commonly supercedes ordinary biological fitness as our goal (Flanagan 2002; chapters 1 and 2 above). There is a dispositional account of fitness that might say this: your average Belgian person is more fit than your average person from Bangladesh because he or

she is generally in better health, etc., so *if* they were trying to reproduce at the same rate the Belgian *would* produce more viable offspring.

35. Simple vertebrates first appeared during the Cambrian explosion circa 500 million years ago.

36. I believe the concept of omniscience is exclusively Mahayanan. In fact, it is one of the reasons why Tibetans think less of "Buddhas by way of Theravada." Dharamakirti (seventh century) clearly thought that omniscience was a central concept.

37. My diagnosis of what is going on here is this: The idea of rebirth is not played up much in the Dalai Lama's book, although it is mentioned several times. We know enough about Buddhism to know that the idea matters. And the normal way it is described and plays out is as follows: *Karmic causation* does not in fact operate in the tame way any naturalist armed with the combined resources of the natural and human sciences can accept and potentially explain. Karmic causation is not simply an innocent way of referring to mental causation and what it gives rise to, to the sum of all good and bad effects that humans create and that affect future generations. These effects are subject to a complex payoff system that involves a cycle of conscious rebirth. If I live well, then I live well in my next appearance. Because Buddhism is *atman*-less, how and in what form "my consciousness" lives again is quite mysterious. Hinduism and Christianity have less trouble making sense of the idea of an afterlife that is both mine and that is, in addition, a sensible reward or punishment for my *karma*. Why? Because they hold respectively that each person is possessed of an immutable *atman* or soul. Despite, many other problems, such views have a much easier time than Buddhism which rejects the reification involved in such posits as the self, soul, or *atman*, when it comes to explaining what is it about "me" that is reincarnated or that gets to sit at the right hand of God. It is "my" essence, "my soul," "my Self." In order to keep things simple here I am not going to delve more deeply than necessary into the thorny issue of rebirth. I devote appendix 1 (at the end of this chapter) to a more complete discussion of *Buddhism and Rebirth*. For the very curious, appendix 2 is devoted to a discussion of the Roman Catholic position on evolution—there are some interesting similarities and differences.

38. One exception—with a growing number of fans—comes from work on self-regulating complex dynamic systems. Here there is a reintroduction (maybe) of a type of Aristotelian formal and final causation, but in immanent temporal space (without the future's actually causing what happens how). The basic idea, eloquently expressed by Kauffman (1995), involves excellent explanations and predictions that can be made about the behavior of certain complex systems if we posit a certain directionality governed by "attractors." Most evolutionary biologists think that evolution has built emergent phenomena of increasing complexity. Complexity theory helps explain how.

39. Buddhists might take comfort in the fact that contemporary physics, especially string theory and its close cousins, is almost completely mathematical, inferential,

and far as most eyes can see, not testable! At a Mind and Life meeting in Washington in November 2005, Father Thomas Keating remarked "Nowadays you'll hear many, many more wild and crazy ideas about the nature of things in the halls of physics than in any Sunday religious service." True.

40. Thus, Nagarjuna's doctrine of emptiness (*sunyata*) is rightly read as a phenomenological response to a certain psychic atomism that members of the Madyimaka school see remaining in the descriptions of the original *Abidhamma*. 'Emptiness' is the thesis that things (such as the mind or the self) that seem to have an essence don't. Pre-Buddhist psychology decomposed mind into familiar sensory and cognitive faculties. Nagarjuna claims to see that these faculties admit to further phenomenological decomposition, as do whatever elements seem to make up the next lower level, and so on, *ad infinitum*.

41. When two lovers meet, they will normally, at some point, use the language of love: "I love you," "I love you too." But it is a familiar fact that what was said to be love was for one or both parties infatuation and/or lust. It can be hard to know first-personally which it is, and there are social pressures to speak the language of love. But this language is hard to learn and teach because we are talking about individuating closely related inner states. When distinctions are required for external objects (e.g., a baby's juice cup, a teacup, and a coffee cup), the community of external observers is positioned to catch and correct more easily mistaken verbal behavior.

42. The reason behind the inference is, of course, not a straight shot from phenomenology to ontology. There is also the 2,500 years of development of Buddhist philosophy and "religion" that are only partly dependent of phenomenology. There is the metaphysics of karmic causation, rebirth, *anatman*, emptiness (*sunyata*), the myriad views about what *nirvana* is/consists of, and so on.

43. My argument does not depend on success in the project of vindicating type neurophysicalism. What matters for me is that mind or mentality be realized physically (token neurophysicalism). Some naturalists think that if only token neurophysicalism is true then consciousness may not be a natural kind that serves any useful scientific purposes. Dennett compares the situation to "health." Health is not supernatural, but it is not locatable in a single physical system. Second, Koch's work, which began 15 years ago in collaboration with the late Francis Crick, is the most philosophically sophisticated and scientifically responsible work on consciousness yet written. Koch, being humble, would be the first to admit—in a Newton-like way—that he has been helped to see as deeply as he does by the work of various under-laborers from James to the present. In the mid 1990s I had the amazing opportunity to spend the better part of a week discussing consciousness with Koch and Crick at Caltech.

44. Koch adds the following judicious thoughts about the exact relation between first personal experience and the brain: "Along the way, the great debate that swirls around the question of the exact relationship between neuronal and mental events

needs to be resolved. Physicalism asserts that the two are identical; that the NCC for the percept of purple *is* the percept. Nothing else is needed. While the former is measured by microelectrodes, the latter is experienced by brains. A favorite analogy is with temperature of a gas and the average kinetic energy of the gas molecules. Temperature is a macroscopic variable that is recorded by a thermometer, while the kinetic energy is a microscopic variable that requires quite a different set of tools to study. Yet the two are identical. Even though, superficially, they appear quite distinct, temperature is equivalent to the average kinetic energy of molecules. The faster the molecules move, the higher the temperature. It does not make sense to talk of the rapid molecular motion causing temperature as if one is the cause and the other the effect. One is sufficient and necessary for the other.... At this point, I am not sure whether this sort of strong identity holds for the NCC and the associated percept. Are they really one and the same thing, viewed from different perspectives? The characters of brain states and of phenomenal states appear too different to be completely reducible to each other. I suspect that the relationship is more complex than traditionally envisioned. For now, it is best to keep an open mind on this matter and to concentrate on identifying the correlates of consciousness in the brain." (2004, pp. 18, 19)

45. The full explanation of how methods of self-cultivation, as opposed to the explanation how conscious experiences are realized and how their causal efficacy is possible, will have to go wider than just what happens in the brain. History, ethics conceived as human ecology, eudaimonics, sociology, the actual tradition in which one is embedded, etc., will be required to explain why individuals or some group of individuals find a set of "spiritual" or mind-training practices attractive, as productive of flourishing. Here we go into the historical and contemporary environment to explain what an individual experiences and why. But the environment is conceived in fully naturalistic terms (no miracles, no supernatural forces, and no immaterial forces are required or allowed). One more point worth emphasizing: There are certain "experiences" such as the "skin and spine tingling thrill" of hearing beautiful music, or of sexual experience, which Gillian Einstein and I have suggested are best understood as "whole body experiences" and thus as requiring "whole body explanations." Thus, for example, experiencing one's current sexual feeling as full-steam-ahead lust, or experiencing one's general sexual orientation as lesbian, gay, or heterosexual involves bringing in hormones, and the like, to the explanation (or constitution) of the experience. In such cases, looking to brains as the sole place where conscious experiences occur may be too limited (Einstein and Flanagan 2003).

46. Obeyeseke thinks that the pressures to develop non-karmic eschatologies are powerful enough to think they existed on the Indian subcontinent in small villages long before the Vedas appeared.

47. There is the idea that the ordinary people won't behave without some such high-stakes narrative in place. Some do think this. I don't. Hope springs eternal and all that. See the discussion of the Bachelor-Thurman debate in note 5 above.

48. The logic of 'ought' and 'can' are relevant here. Many think both in ethics and epistemology that "ought implies can." No one thinks "can implies ought." In this situation under discussion this problem arises: You ought to accept neo-Darwinian theory. If you want to be a bone fide Catholic in good standing, and so on, you can't. It is not permitted. If you are given the choice you should do what is most rational: believe in Darwinism, and reject the beliefs that are incompatible with it.

Chapter 4

1. There are good generalizations in mind science, although the quality and precision of the generalization depend on what domain one looks at. The visual system is the best-understood of the five sensory modalities. Because there are 20+ areas of visual cortex involved in "seeing," the laws about how normal processing occurs are functional and system-level. There are many excellent generalizations about, for example, how shape detection is "done" here and color detection is done "there" and how they bind unconsciously. There are no laws like $f = ma$, and the consensus now is that as in medicine there never will be such laws. Why? Not because visual scientists are not smart enough, but because the biological system in question is a complex functional system that behaves in a normal and regular manner, but that admits of irregular well-functioning designs; and this is so because that is the nature of biological things. If a person has a stroke that affects a particular aspect of visual processing, say motion detection, neurologists know where the damage will show up and do a "look-see" to confirm it. But because each brain is slightly different at each and every level, the damage may be sited slightly differently in two individuals who show exactly the same deficit.

2. Becker (1998) shows that, and how, a noble Stoicism is still a live option.

3. In various ways these philosophers retained admirers until the nineteenth century. Descartes, Spinoza, Adam Smith, Kant, Nietzsche, and Marx all engaged the work of one or more of these philosophers. And the American Founding Fathers who had read Plutarch's *Moralia*, which includes his famous "Lives" of the major Hellenistic philosophers, as well as in all likelihood Cicero and Seneca, reveal that influence in well-known ways in the "Declaration" and the "Constitution."

4. There is controversy about whether and if so how the Greek and Indian civilizations interacted. Alexander the Great's soldiers entered, and many remained in, India. Furthermore, there may well have been traffic in both directions prior to 1500 B.C.E. (See McEvilley 2002 and Sutin 2006.)

5. Confucianism also has a long-tradition of promoting self-cultivation techniques. See Ivanhoe 2000; Slingerland 2003; Shun and Wong 2004.

6. *Abhidhamma* is Pali; in Sanskrit it is *Abidharma*. Normally I use Sanskrit terms rather than Pali ones because they are more familiar to Western eyes. Here I reverse

the procedure and use Pali rather than Sanskrit throughout since the *Abhidhamma* is composed in Pali. Furthermore, it is a sign of respect for the tradition, since the historical Buddha who lived around 500 B.C.E. is thought to have spoken in Pali, not Sanskrit. The first texts, the three baskets of the Pali canon are composed/written down around 100 C.E. The *Abhidhamma* consists of cosmology, metaphysics, and psychology. The prefix 'abhi' means, or suggests, the drawing of distinctions. It is attached to 'dhamma' (Pali; in Sanskrit it is 'dharma'), which in this context refers to the teachings about the bare ontology of things. Books 1 and 7 are the main ones devoted to psychology, the other books are devoted mostly to Buddhist views on time, causation, cosmology, and so on. The other two baskets of Buddhist wisdom contain parables (Pali *suttas*, Sanskrit *sutras*) and moral rules for monks, nuns, and the laity.

7. Julia Annas has warned me about assimilating Buddhism to a eudaimonistic theory. There are deep pragmatic and consequentialistic aspects of Buddhist ethics. That said, Buddhism does contain a view about what the necessary conditions for flourishing are. On Buddhist ethics as powerfully consequentialist, see Goodman 2007.

8. This deep respect for the *Abhidhamma* obtains despite the fact that many Mahayana Buddhists see the original manual, penned by Theravadan monks, as too glowing in its treatment of the monastic life, as well as still embracing remnants of the doctrine of *atman*, as well as not having really absorbed the idea that at bottom all things are empty (*sunyata*). The complaint is that the decomposition of mental states bottoms out in, what seem to be, indestructible psychological atoms which themselves should be considered conditioned (because of the *doctrine of dependent origination* = everything is in flux, dependent for its ephemeral being of prior conditions), and thus that are further decomposable *ad infinitum*.

9. *Citta* and the *cittas* are analytically distinguished from the mental factors (*citasekas*) that they, as it were, can contain. So, roughly, Joy-Consciousness might contain joy-at-an-infant's-birth-in-my-family or joy-at-a-friend's-success. Joy-Consciousness is a type of consciousness, thus a *citta* of *Citta,* whereas joy-consciousness-about-family and joy-consciousness-about-friends would be two sub-types (factor, *citaseka*). Even the *citasekas* admit of lower-level distinctions that happily are not made! But to give a feel: I might be happy [that sister Nancy has a baby] and [that sister Kathleen has one]. The feeling is of the same type, but the intentional content, marked off by brackets, differs in the two cases.

10. Similarly, one might feel happy about one's friend's successes ("sympathetic joy") but have failed to notice that the successes were not achieved in an honest way. Whether one would be judged culpable for this sort of ignorance depends on what was in view and what wasn't. The point is that certain epistemic deficiencies can undermine the warrant, and thus the sublimity, of being in an (otherwise) divine state of mind.

11. David Wong wonders if this isn't excessively moralistic: "Why not just say a state of euphoria caused by a seizure is neutral?" Agreed.

12. In classical Buddhist sources, there are three separate disciplines for the study of consciousness. The *Abhidhamma* focuses on examination of the causal processes of the hundreds of mental and emotional states, our subjective experience of these states, and their effects on our thoughts and behavior. It is related to what could be called psychology, including cognitive therapy and phenomenology. Second, Buddhist epistemology analyzes the nature and characteristics of perception, knowledge, and the relationship between language and thought in order to develop a conceptual framework for understanding the various aspects of consciousness—thoughts, emotions, and so on. Finally, Vajrayana uses visualization, thoughts, emotions, and various physical techniques such as yogic exercises in an intense meditative effort to accentuate wholesome ways of being and to transmute the afflictions of the mind. It is concerned not with discovering an independent permanent entity called "the mind" but rather with understanding the nature of the ordinary mind and effecting its transformation into a non-afflicted, clearer state.

13. These two quotations can be read as making different claims—the first about what produces "happiness" and the second about what makes for a "meaningful life"—but this doesn't matter for present purposes.

14. I use the plural because, as I said in the preceding note, these statements make somewhat different claims at least on the surface. The first is a claim about what brings happiness. The second about what brings a sense of meaning and purpose with happiness possibly added as "gravy." Anyone interested in testing these two hypotheses would need to unpack them further.

15. My internalist objection is about justification, and has nothing to do with the debate about internalism and externalism as it relates to moral motivation.

16. Aristotle, of course, did not know the law in this form.

17. Jonathan Haidt (personal communication) explains that happy people make more money. In all likelihood this is because they are happy, appealing, extraverted, and so on. This is what is called a "reverse correlation." Two other points from Haidt: (1) Comparative economic advantage has some definite effect on first-person and third-person judgments of subjective well-being. (2) Some data show that money might be (for more than a few) about as important a causal contributor to subjective well-being as the number and quality of close relationships. A final observation: Wealth and health go together. Health, especially when it comes to longevity, has a strong relation to stress levels (cortisol is the nasty culprit), so it may well be that wealth allows people to escape certain statistically normal stressors, such as not having enough to pay the bills.

18. In general, I recommend not accepting the "immaturity" charge. Those who say such things as that "psychology as a science began in Leipzig in 1849" are trying to

remind us of psychology's youth. But let's face it: "Brass instrument psychology" may begin in the late nineteenth century, but deep psychological inquiry is as old as every great wisdom text. Overall, I think it best to see many of the problems with mind science, as well as with all *Geisteswissenschaften*, as having to do with the epistemic difficulties of studying very complex dynamic systems. Add to this the fact that one cannot take the objects of inquiry—persons—out of the complex world and into the sort of constrained experimental situations that we use to study some other complex systems.

19. Hume was keenly aware of clerics who tried to pull the wool over peoples' eyes by seeming to derive categorical moral prohibitions from alleged facts such as that in the Bible God says such and such and therefore you ought not ever do thus and so. That God said such and such is, if he said it, a fact, a description of something that happened. Nothing normative deductively follows from it about what you ought to believe or do. Nada. Zero. Well you can keep repeating it. If p, then p.

20. I don't want to make any general claim about progress in science, technology, the arts, the crafts, or ethics. In the 1960s there were many shoemakers who would put an extra quarter-inch of heel on a shoe. My Dad needed this. So do I. But now such shoemakers are hard to find. So too certain virtues, such as honor and loyalty within one's community, might get watered down, even lost, as we move from small agrarian communities to industrial urban life.

21. Even when one sees such "universals," there are always questions about who the "innocents" are and what kind of property "rights" are acknowledged—food produced by my hands, my slaves, my wife?

22. It is useless to think the request is being made of Athenian ethics that it be more expansive. That Athens is over. But insofar as we are heirs to that tradition, the request can be made of us. There is this problem that will make simple addition impossible: It seems unlikely that the virtues and the values on the A list and the B list are consistent as conceived from the perspective of each original tradition. The Buddhist might balk at "courage" (too military) or "magnanimity" or "greatsouledness" (only acquisitive people will be positioned to embody them). The Aristotelian might be bewildered by the sort of expansive love for all sentient beings (Spartans) advocated by Buddhists. So lots of nitty-gritty adjustments will be required to achieve consistency. Here, as elsewhere, the procedure of mature wide reflective equilibrium is the only one ever invented to rationally resolve such problems. I say "rationally resolve" because conquest can require accommodation or blending of two ethical traditions. One final point: I am not claiming that the Buddhist wins overall. Aristotle is much better on the role of friendship in leading an excellent human life than is Buddhism.

23. In new work (in progress), Haidt and his colleagues have added an in-group/out-group module. I say nothing directly here about the wisdom of adding that module. But Haidt wisely never claimed completeness for his foursome; nor do I

assume it. Some things I do say below might make one wonder whether in-group/out-group might not be an adaptive problem domain rather than a module. I reserve judgment.

24. See note 23 to the present chapter.

25. Haidt and Joseph rightly cite George Lakoff's 1966 book *Moral Politics* as making closely related points. I recall reading Lord Patrick Devlin's *The Enforcement of Morals* in the mid 1970s. Devlin, a very respected liberal British jurist, was asked to respond to the Wolfenden Report, which had argued for decriminalizing consensual male homosexuality on Millian grounds: it didn't harm anyone. Perhaps to his own surprise, Devlin recommended caution: If a practice—he thought homosexuality fit the bill in the UK in the late 1950s—provokes strong feelings of "intolerance, indignation, and disgust," don't legally permit it. *Rationale*: distinguish between the "reasonable person" and the "rational person," where the first is the unreflective ordinary person and the latter is the reflective person who might be able to produce the argument to the effect that a practice does no real harm. The moral fabric of society is held together by what "reasonable people" think, not what the handful of "rational people" think. Haidt's model doesn't require endorsing Devlin's conclusion. But it helps make sense of it. One observation on Devlin: he seems to think a practice like homosexuality would have aroused the disgust response in virtually everyone at the time he wrote. But for this to be true, the homosexual population, whatever size it is—should have the disgust response too. This is possible, I guess, if even gay people self-ascribe disgust to what they are. But the rational person will, I think, be able to explain why this is sad, the result of lack of compassion, as well as due to false views (possibly self-ascribed) about the choice involved in sexual orientation.

26. Haidt does not endorse the idea that the universal virtues cited are, in fact, universal. Even Peterson and Seligman sometimes use the language of ubiquity rather than universality. Haidt's reasoning, which I endorse, is this: the intuitive modules are universal; some very few adaptive problems (needing to stay alive) might also be universal. Beyond that what norms, values, and virtues evolve will be variable, based on mixes of who has power, what local conditions demand, what resources for problem solving exist, past historical solutions. Everything interesting will be in detailed philosophical-anthropological analysis.

27. On the first day of ethics class, I tell me students about a marriage practice that has been common for 2,000 years among nomads in Nepal—at times as many as 25 percent of Nepalese engage in the practice: a daughter marries all the brothers in a family. American students find this form of polyandry disgusting. I probe. It makes the women feel that something is amiss because of what seems like infidelity, as well as because of prospects for sexual exhaustion; the young men simply hate the thought of sharing their wife, especially with their brothers. A different example: Suppose American college students started coming to class "topless" in the spring. I would find this distracting, but not disgusting. It would be unconventional, and if it

just started with no warning—possibly a sign of bad manners, but I, at least, wouldn't moralize it. At a nude beach it is not even distracting.

28. I will not take up the question of the status of the social intuitionist modules as modules except to say this much: Haidt acknowledges that his modules are not *special-purpose* in the way the five sensory-perceptual modules are. Mistreatment of my sister (injustice) will be perceived as such and arouse disgust and contempt. If, at the same time, I see you as unworthy slime relative to my clan, hierarchy intuitions/judgments will be in play. The point is that a single situation can activate (result in the interpenetration of) several or even all of the social intuitionist modules at once.

29. If it hasn't crossed your mind by now, I will offer you a basis for objecting to my formulation of the standard for "objective flourishing": What I have offered is, at best, something like a set of standards based on rational *intersubjective* agreement. True. But such agreement is answerable to a wide set of facts—the nature of persons, local conditions, power maneuvers, and predictions and other forms of evidence about what produces eudaimonia. Furthermore, rational intersubjective agreement is how all the sciences operate, even those in which practitioners use the more honorific language of "objectivity" with the connotation that they are speaking about something more that "rational intersubjective agreement." That said, this seems true: the external world would be as it is even if we were not present. Values depend on us. If values (moral, aesthetic, etc.) exist, they need persons to do so. This is true. Two points: (1) Colors are creature dependent. (2) Given that there are sentient beings (persons plus), there are better and worse methods (and reasons) for flourishing.

30. Jeremy Evans points out that Churchland could help his argument by deploying an interesting and important argument from Robert Wright (2000). Wright argues that some moral strategies (or skills) will naturally contribute to more productive societies, which will then replace less productive ones. (This applies mainly to big-ticket moral items, including property rights, rules against violence, reciprocity, and promise-keeping. The vagaries of mundane tax laws may not contribute to a more flourishing society.) Wright's argument is that if people simply follow their own self-interest as they play "non-zero-sum games," there exists an equilibrium, which will tend to protect the rights of property and life. In short, he believes that the direction of history is that the world is becoming a more efficient and safe place to play "non-zero-sum games" Cultures that find this equilibrium are going to flourish at the expense of the ones that do not—they will have more flourishing economies and expansive trade routes. Of course, trade routes are only helpful if one can assume the safe passage of goods and people. Therefore, trading societies set up rules to protect people and property, so they can continue to play those non-zero-sum games. So when the supposedly ruthless Mongols invaded China, they created a society of peace along roads that had long been treacherous and dangerous. Safety was not motivated by the Mongols' kind and loving ways but by their own self-interest as

they continued to expand their abilities to play non-zero-sum games. So over the long haul we should expect to see the propagation of non-zero-sum games, and that process will lead to safer (perhaps read moral) societies. In this sense, we can support long-term positive—i.e., progressive—directionality to moral history, even though for any given segment of history we examine we may not be able to predict whether there is progress.

31. By 'independent' I mean something like this: the new commercial norms govern only trade with the new group, but don't pertain to the original group's commercial practices with each other. There are many current examples of trade agreements of this form.

32. It is curious that Rawls thought that Americans will have the "right intuitions" to support his theory of justice, since few Americans seem attracted to the "Difference Principle." Northern European welfare states in fact embody some form of the "Difference Principle," so I could see that principle having intuitive appeal there. Many Americans read the principle, correctly, as socialist, and American intuitions about desert (from Locke et al.) hardly favor it.

33. Hagop Sarkissian points this out to me: "You seem to use 'intuitions' in the sense of innate 'reactive-attitudes,' but Damasio seems to use 'intuitions' interchangeably with 'somatic markers', which are forged through learning. If so, then what we have is not, strictly speaking, a continual exercise of gaining rational foothold over these base intuitions, but rather of continually reforming them." This is a really important point, which I need to think about. First pass: I may be unwise to adopt without comment Haidt's use of 'intuition' and 'intuitive'. I normally read him as using these terms in a way that could cover the initial settings of the reactive attitudes as well as the way they, still considered as reactive attitudes, are modified by learning. The fact is that Haidt usually sees the relevant action in moral response as occurring in the emotional regions of the brain. Damasio's somatic marker hypothesis allows that learning moral habits is more systemic, involving the pre-frontal cortex as well as, say, the amygdala.

34. The insight here is ancient: One sees it throughout Confucius's *Analects* when he worries about the effects of "village worthies," charismatic leaders, teachers and so on who lack *ren*, *li*, and *de*—humaneness, respect for the rites, and virtue. The "village worthies" may seem to possess virtue but they don't, and as a result, the citizens and young people are misguided in their socio-moral-political development. Mozi, to my mind the most underestimated of the classical Chinese philosophers, sees this problem clearly and accuses Confucius of not seeing how serious this problem is.

35. Compare Reichenbach's pragmatic justification of induction: It is impossible to prove that induction *will*, in fact, work; what we can show is that *if* any method can produce knowledge, *it* will.

36. See Josh Greene's homepage: http://www.wjh.harvard.edu/~jgreene/.

37. Introducing such principles will of course work best if one is already a fallibilist and a naturalist. Some normative systems make the mistake of thinking they have everything right and, because they are based on divine instruction are unrevisable. There is some historical evidence that external pressure can sometimes get such beliefs about, say, literal interpretation to yield. But it *is* a serious problem.

Chapter 5

1. Associated Press, Reuters, the BBC, and Canadian and Australian Public Radio were the first out of the gate with reports on the research, and I did (too) many media interviews. *Dharma Life*, in an amusing headline of its own, called the scientists who performed the early studies on the meditating monk "joy detectives."

2. Matthieu Ricard (2006) calls happiness life's "most important skill." The idea that happiness is a skill rather than a state goes back to Aristotle. Aristotle is often interpreted as suggesting that happiness comes from skillful activity (in which case it could be interpreted as a state), but he is just as reasonably read as suggesting that what he calls 'eudaimonia' *is* the skillful activity of living well.

3. "Mattering" is of course a normative notion. The fact that people apply discount rates to backaches but not to loving sexual encounters may not be surprising, or indefensible, but it does seem to employ criteria of "mattering." But psychologists worry about applying discounts. The self-serving bias (SSB)—so dubbed because people report, and apparently remember, good things they do better than bad things they do—is worrisome. In contrast with the backache–sexual experience cases described, the SSB does, on most any interpretation, involve self-serving spin and possibly not taking sufficient responsibility for actions one should be taking account of. Thus, if we switch the example to two sexual encounters over, say, a two-day period, one licit and the other illicit, the latter is likely to be discounted and this might easily be construed as problematic since from a moral point of view we might be inclined to say it *should* matter.

4. There is the issue of what in fact matters most to most people and the issue of what *should* matter, where what *should* matter may in fact matter to hardly anyone.

5. In epistemology it has long been common to think that instantaneous state reports are much more reliable than all-things-considered reports over an interval. Descartes' "cogito" can be read as an instant self-confirming state report. In a different conceptual space, the logical positivists loved "protocol sentences," such as "Lo, red-patch there-now," thinking such reports were the observational rock bottom for good science.

6. Put these issues to one side. There are still some problems with the measurement tools. SWB measures typically specify some set of domains $(d_1, d_2, \ldots, d_n)$ as relevant

to judgments of SWB, but sometimes only ask for an overall score, rather than for scores for each *d*. Other times scores for each member of the set $D = (d_1, d_2, \ldots, d_n)$ are given, as well as an overall score. But as I read the literature, the overall satisfaction score is still normally given by the subject. That is, the overall "life satisfaction score" is not derived by weighting and then performing a computation over scores assigned to each domain. More usually, the subject gets to assign all the scores, first the overall life satisfaction score, and then the domain scores. The only exception I see consistently (and mostly in recent literature) is one where subjects give an overall score for the amount of pleasant experience and for the amount of unpleasant ones; the latter is then subtracted from the former, and an overall "hedonic affect" score is computed.

7. According to table 5.2, Calcutta sidewalk dwellers give a very low score to overall life satisfaction and give significantly higher (but still low) scores to all the other domains. What is going on? What is such an overall score even about? Is overall life satisfaction a domain among others, or is it a score based on computing the domain scores, or is it a separate score that subjects are asked to give. There should be a clear answer but I haven't found it in the literature. My surmise is that the subjects give all the scores, that is, as I noted above the "overall life satisfaction" score is not computed from scores in the domains. So assume that the Calcutta street dwellers were asked to give an overall life satisfaction score. Even so, they might not have done that. That is, these individuals might not really be giving an overall satisfaction with life score but, instead be focusing on what it is like to live on the street while discounting the specific domains—all of which are better. i.e., more satisfactory. It seems to them as if life sucks overall, but if they were to reflect on the domains that matter to them (first, say), they'd see that it sucks less than they are inclined to say without being asked to focus domain by domain.

8. See, for example, Diener and Scollon 2003. One might regiment the overall score by determining a mathematical function that assigns different values to the different domains—either as assigned by subjects themselves (so family is weighted more than work for some individuals, the reverse for others), or by psychologists who find that some domain, family, say (almost) always outweighs "work" by some characteristic amount or ratio. What I have in mind here is that even in cases where satisfaction in the two domains of family and work receive identical satisfaction scores, say +7; we (either the subject or the psychologists) might weight the first 0.6 and the second 0.4 in determining SWB on the basis of how much importance members of the population assign to these domains.

9. Social reinforcement is so important that it is hard to imagine that the person I describe can keep her eye on virtue for herself and her charges for two generations despite the lack of social support. Further, the *capabilities approach* of Sen and Nussbaum, discussed in chapters 1 and 2, is an approach designed in part to allow us to make these sorts of objective judgments of flourishing or non-flourishing that (can) diverge from SWB reports or assessments.

10. This doesn't yet measure how well they are in fact doing in relation to the norms, values, and domains designated by PWB.

11. See also Flanagan 1996a, 2000b.

12. The program of neurophenomenology advanced by Varela and his colleagues is virtually identical to what P. S. Churchland (1986) calls the "co-evolutionary strategy" and what I (1992) call "the natural method." A great advance being developed by Varela et al. just before his untimely death was second-person phenomenology. The basic idea was to have experienced individuals work in a cooperative Buberian I-Thou manner, with subjects offering first-person reports to help them get clearer on what they are experiencing in fine-grained detail. The idea is to improve first-person reporting by a form of dialogue "between the two of us." The subsequent, hopefully more refined report, is something the subject and her interlocutor work to produce together.

13. Gary Schwartz and Cliff Saron were also involved at early stages.

14. To this day, the best book by a philosopher on the phenomenological structure (somewhat idealized) of intentions and plans is Bratman 1987. Before that, in psychology, there was Miller, Galanter, and Pribram 1960. In retrospect, we can say that both books lay out from, what Dennett calls the *intentional* and *design* stances what is in fact executed, from the point of view of the *physical stance* by PFC.

15. Deficit studies are terrific ways to figure out how a complex system works, so looking at areas affected by stroke, tumors, etc., has been very helpful in distinguishing among areas differentially involved in particular types of processing. PLFC strokes are correlated with increased depression, but are neither necessary nor sufficient causes of stroke related depression.

16. Imagine that, as is often the case, I have one and the same text—say, this very chapter—on different computers, a PC and a Macintosh, with different operating systems. Each computer holds, realizes, encodes, and projects the identical text when asked to do so, but each does it in a somewhat different way. The *identical* text on the screen is sometimes referred to as a "user illusion" since behind the scenes differences abound.

17. Some right-leaning PFCs might be indicative of happiness. In some people, most often some subset of left-handed people, hemispheric differences are reversed.

18. My considered view, which I violate somewhat here, is it that it is best to keep the word 'happiness' in the titles of books, chapters, or lectures—it brings in buyers and audiences. But the word, not the concept, is best kept out of the analysis. Here I follow Marty Seligman's strategy. Seligman is the *eminence gris* (actually he is on the bald side) of the positive-psychology movement. He entitles his best-selling book on positive psychology, *Authentic Happiness* (2003), and then rarely uses 'happy' or 'happiness' in the text. We are then told this in the appendix: "The word *happiness* is the

overarching term that describes the whole panoply of goals of Positive Psychology. The word itself is not a term in the theory.... Happiness, as a term, is like *cognition* in the field of cognitive psychology or *learning* within learning theory. These terms just name a field, but they play no role in the theories within the field." (2003, p. 304)

19. The idea is that the basal ratio is somehow deep and abiding (deep-structural) for an individual. But presentation of "good" or "bad news" results in tonic shifts. These can be thought of as (normally) temporary and involve surface structure. A phenomenon known as "adaptation" where lottery winners or those who suffer loss of loved ones or become paraplegics "return to baseline" after a surprising short interval. Such individuals might be thought to be evidence of the resiliency of the basal deep-structural features of LPFC:RPFC ratios. But there are other explanations for adaptation. It is a topic that needs more bright minds to get to the bottom of the phenomenon.

20. I am not sure why my article got so much media attention, it has to do I think with what the AP and Reuters decide to write about, because I was not the first to report the findings on Matthieu. The Dalai Lama first alluded to this study in an op-ed piece for the *New York Times* (Gyatso 2003a). I reported on the study in *New Scientist* (2003a), and Dan Goleman (2003b) wrote about it in the *New York Times*.

21. These facts put a burden on scientists in the field: When a worthy sort of happiness is claimed to be among the goods produced by embodying a practice, two requirements must be met by scientists wishing to study the connection between that practice and that kind of happiness. First, we must specify precisely what kind of happiness we are looking for. Second, we must have a clear conception of what aspects of the practice are thought to be the main contributors to attaining that kind of happiness.

22. I use 'enlightenment' or 'awakening' (*bodhi*) and 'wisdom' (*prajna*) interchangeably—often as enlightenment/wisdom. Strictly speaking, they are not exact synonyms. But using them interchangeably is quite common in the literature, and comports with one classical view according to which achievement of complete wisdom and complete virtue are necessary and sufficient for enlightenment (*bodhi*). For parallelism, I use virtue/goodness, or just one or the other, to refer to a life of good conduct (*sila*), as well as a character that embodies eventually, *inter alia*, the divine illimitables or abodes (*Brahma-vihara*), compassion, loving-kindness, sympathetic joy, and equanimity.

23. It is only by living a life of wisdom and virtue/goodness that a sense of meaning, purpose, and happiness can be secured. It is a Zen-like paradox that if we seek simply to attain happiness we won't, whereas if we aim for enlightenment and virtue, initially setting our undisciplined pursuit of happiness to the side, that we might begin to achieve "true happiness"—happiness^{Buddha}. Happiness^{Buddha} is the kind of happiness that comes from commitment to and embodiment of Buddhist philosophy

understood as a normative theory and practical philosophy of enlightenment or awakening and virtue or goodness. It is a type of happiness born of achieving wisdom (*prajna*) by becoming free of the standard mental affliction that come with being human, and finding one's way to deep compassion (*karuna*) and loving-kindness (*maitri*) for all sentient beings. Although I have claimed that enlightenment and virtue co-constitute the *ultimate end*, we can analytically separate the two components. One component is *Enlightenment/Wisdom*: Buddhist enlightenment requires that one comes to understand: (1) that all things are impermanent (*anitya*; Pali, *anicca*); and (2) for this reason I am not possessed of a permanent self, ego, or soul (*atman*). "I" am *anatman* (Pali: *anatta*), a transient "being" constituted only by certain ever-changing relations of psychological continuity and connectedness (MN.1.138; Siderits 2003, 2007). Another component is *Virtue/Goodness*: Buddhist virtue/goodness requires moral conduct (*sila*) and thus conformity to the third, fourth and fifth of the steps on the noble eightfold path. True virtue, of course, requires more than moral conduct. An individual, such as a bodhisattva, overcomes the three poisons of greed (*lobha* or *raga*), hatred (*dosa*), and delusion (*moha*), and positions herself to embody the four divine illimitables—compassion, (*karuna*), loving-kindness (*maitri*), empathic joy (*mudita*) and equanimity (*upeksa*) (Shantideva, eighth century C.E.; Lopez 1988).

24. See the important study by Davidson's group (Lutz et al. 2004) in the prestigious *Proceedings of the National Academy of Science*. This work shows that there are significant and unusual oscillatory gamma-wave patterns in the brains of experienced meditators compared to controls. Now we need to figure out what these differences mean and exactly what sorts of experiential differences they subserve.

25. When Dan Goleman and I first reported the face reading studies, the word was that no one had ever done as well as the adepts. This turns out to be false. There were a few others in the big database Paul Ekman possessed who achieved similar scores. There is no way to know whether these individuals were skillful face-readers or just lucky. That aside, Ekman has now developed techniques (I have the one hour training CD) that trains anyone to be as good as the four adepts at reading micro-expressions. You can purchase it at http://www.paulekman.com if you have the proper certification and $175. What remains very interesting is why, if it is so easy to learn this skill, everyone hasn't done so, after all there is an arm's race to detect liars and cheaters. Ekman (personal communication) has no answer yet. The fact remains that the adepts naturally developed the skill, but not, as far as we know, by consciously trying to do so for faces. My best guess is that what is called insight (Vipassana) meditation where concentration and skills of analytic attention are honed (often primarily on one's own sensations, mental states, etc.) result in good analytic skills in interpersonal situations. Ekman's first surmise was that it might have to do with skills that come from *metta* (loving kindness) meditation where empathy and compassion are honed. We both might be right. But as of now, we just don't know why/how these adepts developed the skill.

26. There are certain shared principles that are espoused and abided across Buddhist sects. Rahula 1959 remains an excellent source on these shared core beliefs. That said, there are also distinctive practices and beliefs that, for example, distinguish Theravada, Mahayana, Tibetan Buddhism (Vajrayana) Zen (Ch'an), Japanese Pure Land Buddhism, and Socially Engaged Buddhism. To some degree, these differences are based on differences in the choice and interpretation of key texts. But they are also due to philosophical interpenetrations, as in the case of Buddhism meeting Daoism and Confucianism in China, resulting in Ch'an (Zen) which, of course, took on a certain Japanese flavor as it migrated East. Buddhism in North America and Europe is not a sect or a coherent tradition. It is a syncretic blend of Buddhisms. Zen Buddhism first caught North American attention in the 1950s among certain members of the "beat" generation. In the last four decades, Socially Engaged Buddhism (Thich Naht Han 1987, 2004) and Tibetan Buddhism (Dalai Lama and Cutler 1998; Dalai Lama 1999) have become at least as influential as Zen. If there is anything distinctive added to the mix by Westerners (both in North America and Europe) it comes from certain secular and naturalistic impulses, so that, for example, the doctrine of rebirth is either rejected, reconceptualized, considered optional, or understood as a quaint but instructive piece of mythology (Bachelor 1997; Flanagan 2006).

27. In all studies, be they designed to test the connections between mindfulness and health or longevity or to examine the connection between Buddhism and happiness, certain guidelines will lead to well-designed experiments that will yield revealing findings, one way or the other. For example, in cases where experienced practitioners are studied, we will want to know which kind of Buddhism they are committed to and what type of happiness, if any, that kind promises. I claim that happiness^{Buddha} as depicted above captures a common core conception shared across all or most forms of Buddhism. However, there are various more nuanced types or subtypes to be depicted and studied if one wishes to examine a specific Buddhist sect. And truth be told, many non-Western Buddhists will truly say that happiness of any sort is *not* the point.

28. In cases where certain Buddhist practices such as Tibetan Buddhist compassion meditation or the Japanese Pure Land practice of calling on Amitabha's presence (the Spirit of Life and Light) by chanting "Nama Amida Butsu"—are extracted from the kind of Buddhism in which they are typically embedded and are then taught to individuals who have no personal commitment to (possible no knowledge of) any form of Buddhism, this will also need to be carefully marked. The reason is that such studies, no matter what they reveal about the efficacy of that practice in producing some good, even if it is some kind of happiness, have no clear relevance to what many think is most important and interesting, namely, what goods do long-term commitment to (a form of) Buddhism produce. The trick is to direct our natural urge to happiness to the right sort of happiness and then to work with reliable methods to achieve it. The right sort of happiness is happiness^{Buddha}. It comes, if it does come, from practices that aim at enlightenment/wisdom and virtue/goodness (see

SN 4.223–9; SN 4.235–237). There are plenty of texts in the Pali canon (the original canonical texts accepted as complete by Theravadins and accepted as basic but extendable by Mahayanans) where the Buddha distinguishes between types of happiness as understood and chased after by the common person, the happiness of the path to liberation, and the happiness of liberation. He asks: "[W]hat bhikkhus is happiness more spiritual than the spiritual? When a bhikkhu whose taints are destroyed reviews his mind liberated from lust, liberated from hatred, liberated from delusion, there arises happiness. This is called happiness more spiritual than spiritual." (SN 4.235–237, pp. 1283, 1284) Thus, not surprisingly, stories of "happy" *arahants* and "happy" *bodhisattvas* are abundant across various classical traditions (Bond 1988; Lopez 1988). See also Hammalawa Saddhatissa (1970, 1997, 2003), written from a Theravadan perspective. Here Saddhatissa consistently refers to the Buddha as "the Happy One." But he points the reader to *suttas* in the Pali Canon where neither Buddha, nor anyone else, claims to know *direct* techniques for achieving happiness. This, of course, is compatible with there being indirect techniques via wisdom, virtue, and meditation that might produce happiness as an effect.

29. Freud famously doubted that religious illusions would go away. Nietzsche was more hopeful that theistic beliefs would eventually yield to good sense and honesty. Some Buddhist scholars argue that Buddhism makes no sense without beliefs in karma (untamed) and rebirth. Martin Willson (1987, p. 7) explains the situation as follows: "According to the teachings of Atisha and his successors on the Stages of the Path (*Lam rim*), one becomes a Buddhist ... when one takes refuge in the Three Jewels—Buddha, Dharma, and Sangha—with one of three motivations: (1) With horror of rebirth in ill destinies (as an animal, a *preta*, or a hell being) one seeks the aid of the Tree Jewels in order to gain a "happy" rebirth as a human being, *deva* or *asura*. (2) With horror at the suffering inherent in all states of rebirth under the control of karma and defilements (*klesa*), one takes refuge in order to achieve liberation from *samsara*, this cycle of death and involuntary rebirth. (3) With infinite great compassion unable to bear the sufferings of any other samsaric being, one takes Refuge in order to become able to lead them all to liberation and enlightenment, for which purpose oneself attain the powers of a fully enlightened Buddha." Willson adds: "None of these motivations makes sense if one disbelieves in rebirth."

30. See also Taylor 1989.

31. There is lots of interesting recent work on narrative. (See Fireman, McVay, and Flanagan 2003 and Herman 2003.) Narratives are normatively governed: there are norms about not saying embarrassing things, and there are norms about politeness and not saying things that will discomfort self and others. There are also complex norms about giving directional information. By this I mean that supposing that I have learned or am trying to learn from my failures, I give information about my good qualities, worthy accomplishments and plans. Why? Not to deceive myself or others, but to provide reliable information about where I am headed.

32. Several years ago, I saw a television show about a public school in lower Manhattan (Chinatown) where the kids were tops or toward the top in both math and English skills statewide. This was so despite the fact that their first generation immigrant parents, most of who were restaurant workers, spoke Chinese at home. When the principal was interviewed, she was pressed by the interviewer along "You must be sooo proud, sooo happy" lines. To his consternation (I detected), this Chinese-American 50-something woman replied "No, not really, we just try to do what is right."

Chapter 6

1. In the song "Just Like This Train," Joni Mitchell sings of looking for a "strong cat without claws." When I think of spirituality and religion, I have similar desires. Most religions, strong or weak, are like cats with claws that can do harm. Religions, like cats, are the result of a thoroughly human domestication process. No sensible person has a real cat (lion, tiger) as a pet. (I know some people do, but I said "sensible.") Trouble is that many religions treat themselves to a story according to which they are not domestic human creations at all; they are the true, original word of God. Think differently. Tame your religion. Declaw it. Let it be strong and noble. But declaw it. Then and only then can we be assured that it will be a good pet, possibly do more good than harm, provide comfort, and so on.

2. The state of discussion about whether one is a theist, an atheist, or an agnostic is (sadly) informative about the state of religious epistemology. Everyone is an atheist when is comes to most conceptions of divinity. Christians are atheists with respect to Greek and Hindu gods. Hindus are atheists with respect to the Abrahamic God (or Gods). The question of theistic belief makes sense only in relation to a conception of God that is, as it were, on the table for discussion. Most smart people, when they enter this terrain, have not thought about or been the slightest bit impressed by what I have just said. In any case, the dialectic commonly goes this way among open-minded people: agnosticism is respectable but atheism is not. This is insane. There is, let us suppose, a denumerable, but potentially infinite number of conceptions of "a creator." If you ask whether I am a theist or an atheist (or agnostic), you are not, I take it, asking me where I stand on *all* the conceivable contender conceptions; you are asking me about a conception available in and entertained by people in our culture. (Pascal's wager works only, if at all, on a very specific conception of God.) Assuming one does not believe in revelation, because that would be stupid, one ought to be an atheist for each conception of God I have ever been asked, or thought of entertaining. The conception is not normally demonstratively false, i.e., by the standards of deductive logic; but the conception is normally without inductive warrant if we treat them as assertions. So agnosticism is, as best I understand, not an interesting epistemic position with respect to this question, since it treats theistic claims as assertions, as truth functional, but ones where the evidence is insufficient

to decide which assertion to make. But theistic claims are sayings, not assertions, and thus questions about their evidentiary status can't really sensibly arise.

3. In chapter 2, I tried to explain what people who say they are "spiritual but not religious" mean. One might wonder if there are people who are "religious but not spiritual." Although almost no one I know describes themselves this way, there might be some such folk. Such people deny having transcendent experiences or impulses, but believe confidently in following to the letter what they understand to be God's rules for living correctly.

4. The spaces of meaning were created in the first instance from the ground up, by humans living their lives and eventually creating forms of life and institutions that reflect the ways they live and give guidance about worthy things to do and worthy things to be. So in saying that the "desire to connect meaningfully with others is a tentacle that descends from the space of ethics," I do not mean that this desire originated there. It originated on the ground in our social nature. Once the space is articulated however it guides us, on the ground, in this zone of life. This point generalizes across all the spaces.

5. To see how naturalists like me (I am a board member) conceive of our own spirituality, go to http: //www.naturalism.org/. Note that Tom Clark, the director, speaks for many when he claims that a naturalistic picture leads to greater, more universal compassion than traditional theism(s) does. There was a recent meeting in Berkeley in the summer of 2005 to discuss what spiritual, but not conventionally religious folk, can do to bring a spiritual tone to the political discourse on the American left. (For many excellent articles on the response of the left to "theocracy in America," see *Tikkun*, January–February 2006.)

6. I have been claiming that Buddhism is non-theistic. There are many forms, however, especially of Tibetan Buddhism, that tell stories about heaven and hell realms. What is happening? Read such stories as stories that are artfully expressive and motivationally useful for orientating humans toward enlightenment and virtue, but that are not asserted (they can't be) as true.

7. One needs to go outside orthodox Christian circles to discover that at least two of the canonical four gospels were based on a source known as Q, that many early political battles occurred about which gospels would be considered canonical, that the Gospel of Thomas is repeatedly mentioned as not approved, and that the Gospel of Thomas has now been found and is basically like Jefferson's Bible: no virgin birth, no miracles, Jesus is a great moral leader.

8. It seems to me that most people sincerely think that their belief in God is defeasible. Belief in God isn't like a psychotic believing that he is Napoleon. The person with the Napoleon delusion gets feedback that what he believes is true is in fact false, but won't change. Religious believers are not this crazy; they protect their belief from ever getting genuine, disconfirming feedback. If this is right, then theistic asser-

tions are not, strictly speaking, delusions, since they are set up to be immune to any and all countervailing evidence.

9. The whole issue of testability is more complicated than I let on. Many of our best physical theories, or promising ones, are judged by compatibility with the way the world seems to work and in terms of internal consistency. There are no direct tests for many of the ontologically basic stuff posited by elementary particle physics, especially by its string theoretical variant.

10. Art in fact requires a more nuanced analysis than I am able to give in this book. If all art were expressive and not assertive (i.e., not making truth claims), this would explain why there is less conflict (if there is) between art and/or religion in its assertive form and art and science—which is normally assertive, claiming to speak the truth—than there is between assertive theism and science. There is something to this diagnosis. But there are many forms of music, literature, and visual art that activate (and are intended to activate) certain feelings and thoughts that, if activated, lead people to question, for example, religious conservatism or scientistic aspirations to speak adequately about "the sacred depths" of nature and experience (Goodenough 1998). Much folk and folk rock music of the 1960s and the 1970s aimed to speak truthfully about political evil and to motivate people to think and act against such evil. There is also plenty of literature, drama, and film that attempts to speak truthfully about the human condition. Such art depicts the human condition in a different way, in a different language, than empirical psychology, but at the same time attempts to give us insight into the human condition.

11. I want to be clear that nothing in the analysis so far precludes spirituality or religion. It depends on whether the objectionable form of "supernaturalism" is espoused. According to that not-unfamiliar form, of supernaturalism, (i) there exists a supernatural being(s) or power(s) outside the natural world, (ii) this being or power has causal commerce with this world, and (iii) the grounds for belief in *both* the supernatural being *and* its causal commerce cannot be seen, discovered, or inferred by way of any known and reliable epistemic methods. Many forms of naturalistic spirituality reject supernaturalism in the objectionable sense. (See Goodenough 1998.) I myself am spiritual: a Celtic-Catholic-quasi-Buddhist atheist. And I see my ethical commitments as supported and enhanced by deep transcendental cognitive convictions and emotions that powerfully ground a conviction that I am a part of the whole, inextricably connected to everything else that there is. But I reject i–iii.

12. The Oxford English Dictionary suggests that the original *philosophical* meaning of 'naturalism' dates back to the seventeenth century and is "a view of the world, and of man's relation to it, in which only the operation of natural (as opposed to supernatural or spiritual) laws and forces is admitted or assumed." (See Flanagan 2006b and Flanagan, Sarkissian, and Wong 2007.) Barry Stroud (1996) writes: "Naturalism on any reading is opposed to supernaturalism.... By 'supernaturalism' I mean the invocation of an agent or force which somehow stands outside the familiar

natural world and so whose doings cannot be understood as part of it. Most metaphysical systems of the past included some such agent. A naturalist conception of the world would be opposed to all of them." Indeed, Stroud goes on to suggest that anti-supernaturalism is pretty much the only determinate, contentful meaning of 'naturalism'. If he is right, anti-supernaturalism is the common tenet of "naturalism" insofar as "naturalism" is anything like a coherent philosophical doctrine spanning the last four centuries. Let me be clear about a matter of considerable importance: the objectionable form of supernaturalism is one according to which (i) there exists a supernatural being(s) or power(s) outside the natural world, (ii) this being or power has causal commerce with this world, and (iii) the grounds for belief in *both* the supernatural being *and* its causal commerce cannot be seen, discovered, or inferred by way of any known and reliable epistemic methods. Note that the commitment to the dispensability of supernaturalism does not entail a rejection of all forms of spirituality or religion. Theologians and philosophers who are religious naturalists reject the conjunction of i–iii above and that amounts to a rejection of the objectionable form of supernaturalism. Furthermore, many forms of embodied spiritual commitment are themselves committed to the dispensability of supernaturalism in the objectionable sense (i–iii). Buddhism comes to mind, as do some strands in Hinduism, Jainism, many kinds of shamanism, as well as among many liberal Christian communities in America (Quakers, Unitarian Universalism).

13. "Team Dennett" is doing work examining this claim: true believers get divorced less frequently than atheists or agnostics. The evidence so far is that this is false.

14. One caution here: Paul Bloom (2004) thinks that we see in young children powerful "dualistic tendencies." The dualism could be epistemic or metaphysical. Epistemic dualism involves commitment to two different kinds of causation, and two different styles of explanation: (1) ordinary causation and (2) psychological or intentional causation. Metaphysical dualism involves commitment to a world in which there are two ontologically distinct types of substances, events, processes, or properties. Bloom thinks children are natural-born dualists in both senses. I'm not so sure. I grant that all people are epistemic dualists. For many adults this just involves taking the intentional stance and explaining and predicting the behavior of other humans (and more) using psychological language. But many adults (all naturalists) who use both kinds of explanation do not believe in metaphysical dualism. One possibility is that children make no metaphysical divide while using both the language of psychological and ordinary causation. The experimental fact(s) that children are very surprised when a physical object changes trajectory or speed as it enters and exits an occluded space, whereas they are not surprised when a person does so, could mean they just get that people can decide how to move in ways objects can't. If this is true, then becoming committed to metaphysical dualism might require being introduced to a theory, a philosophical theory, specifically a metaphysical one, that is alleged to explain the unusual powers of persons. I do agree that being "natural epis-

temic dualists" primes us nicely to accept metaphysical dualism. The point has direct bearing on whether our natural epistemic dualism directly explains posits of supernatural forces and entities, miracles, and the like. Here again caution suggests this division: Supernatural forces might be conceived as special kinds of occupants of the natural world. The language of 'spirits' simply captures their unusual ways of operating. Many ancient religions, arguably, place their gods in this universe as unusual, but not entirely immaterial, causal players. If so, then the move to immaterial spirits who operate from outside the world involves coming to believe three pieces of theory: (1) that immaterial entities exist, (2) that these immaterial entities live outside this world, but (3) that these immaterial entities, forces, or whatever, are capable of visitations that causally effect what happens in this world. I will assume that we are talking about such a metaphysical theory as I go on to speak about supernatural causes, that is, as = immaterial causes, because that is the view in play nowadays. I am skeptical, at least agnostic, about whether it was always and everywhere the view of the nature of the occupants of the "spirit-world."

15. There is this move: Admit that the principle of conservation of energy is well confirmed, but point out, what is true, no one tracks (no one could, in principle, do so) that the total amount of energy is in fact absolutely constant. But assume it is. If so, it is possible that when some spirit creates some energy here *ex nihilo*, an identical amount of energy leaks out somewhere else in a distant galaxy. Or assume, that the principle of conservation of energy is only well confirmed around here, while physicists assume that nonetheless it applies throughout the universe. Since the principle could be false, assume it is. If the principle of conservation of energy is false, at least not perfectly true, then occasional, even common divine interventions, might not be revealed in any measurements now being done, or even under epistemically ideal conditions. This last point might be falsifiable (if one is being very generous), since it predicts that under much better, or ideal, measurement conditions that the universe would show "extra" energy (assuming "no leakage"). The problem is that the *only* reason to entertain this hypothesis is to make room for miracles. One last point: I often hear people say "That [meaning whatever happened] was God's doing for me what I couldn't do for myself." Or take the Roman Catholic doctrine that God vaginally inserts souls at each conception. (See chapter 3, appendix 2.) One way both these ideas could work is *if* all supernatural commerce with this world is with *immaterial souls*. So God makes me different by making my soul different, or God inserts an immaterial soul "in" each fertilized egg. Naughty idea(s): dualism is a bad idea, worth discarding for logical and evidentiary reasons I and most other philosophers of mind and mind-scientists agree on. Furthermore dualism causes the "theist" one of two problems—possibly both: either once I am changed because my soul is, then assuming the change affects my mind/brain or how I act, we are back to violation of conservation laws; if "soul work" has no effects on the mind/brain or on how I act, then the state of my soul is epiphenomenal; it makes no difference to me, to my life, to who I am.

16. James does not make this mistake. In *The Varieties of Religious Experience* (1892), James takes religious experience as primary and explicitly states that he considers religious institutions and their associated theologies to be secondary. I do, however, have two criticisms of James: (1) He treats spiritual experiences as more common than I think they are, at least independently of systems of practice, "forms of life" designed to produce them, e.g., shamanistic rituals or intensive contemplative practice. (2) He seems to think that religions and religious institutions arise primarily out of such a transcendent experiential base, whereas on his own view he also seems to see the power of non-spiritual factors, politics, and so on, in the construction of religions. On my view, the latter have at least as much power in shaping the institutions of religion as the former, probably more, possibly much more.

17. Meaning and moral glue are also commonly connected to and interwoven with theology, so what I am doing is separating them out for analytic purposes. It matters for someone like myself who would like theology to self-understand its role as myth-making that the meaning and moral glue aspects are, in principle, logically separable from theology—incorrectly understood as a set of epistemically respectable assertions.

18. Humor [playfulness] is depicted as "liking to laugh and tease; bringing smiles to other people; seeing the light side; making (not necessarily telling) jokes." Leaving humor/playfulness out simplifies the discussion. But I also have a principled reason for leaving it out of the discussion. Peterson and Seligman explain that "humor/playfulness" is a late addition that satisfies the universality condition but that was also added because their "classification was too grim without it." This is a very funny reason to add it—almost a joke itself. They say that humor/playfulness may eventually be classified as a "value-added strength"—"most praiseworthy when coupled with one or another strength." I love humor and playfulness; and I prefer spiritual types to not take themselves too seriously, not be dour, etc. But I seriously doubt that there is any interesting connection between actualizing transcendent impulses and being like Charlie Chaplin, Lucille Ball, Bob Newhart, and Bill Cosby (some of their examples).

19. Q: Why don't Calvinists allow married people to have intercourse standing up? A: It might lead to dancing.

20. The Abrahamic religions may be unusual in requiring something close to literal interpretation of their texts. In *The End of Faith* (2003), Sam Harris asks "Suppose a loved one suggests that you give up belief in the God of Abraham and follow the Buddha's path?" Deuteronomy advises: "You must kill him, your hand must strike the first blow in putting him to death and the rest of the people following. You must stone him to death since he has tried to divert you from Yahweh your God." (13: 8–11). The literal interpretation of this text is unambiguous. The moderate must offer a non-literal interpretation and she faces this problem: Deuteronomy also says of the text itself "Whatever I am now commanding you, you must keep

and observe, adding nothing to it, taking nothing away" (13: 1). The Qur'an implores the true believer in Allah this way: "Prophet, make war on the unbelievers and hypocrites and deal rigorously with them. Hell shall be their home: an evil fate." (9: 73, see also 9: 123) Regarding martyrdom, the Qur'an says: "God has given those that fight with their goods and their persons a higher rank that those who stay at home. God has promised all a good reward; but far richer is the recompense for those who fight for Him.... He that leaves his dwelling to fight for God and His apostles and is then overtaken by death shall be rewarded by God.... The unbelievers are your inveterate enemies." (4: 95–101) The New Testament is a kinder and gentler document than either the Old Testament or the Qur'an. But crusaders of many stripes have found support in it for all manner of vicious actions. True believers are forbidden to become Rabbis, to be Jews (Matthew 23: 8). And the demands on the true believer contain all manner of inane observations, advice, and commands. Luke writes: "If any man come to me and hate not his father, and his mother, and wife, and children, and brethren, and sisters, yea and his own life also, he cannot be my disciple."

21. 'Tribalism' is the name sometimes given to communities of narrow ethical perspective. To be fair to actual "tribal peoples," they are often sufficiently isolated that issues of (and thus thoughts about) loving outsiders, let alone everyone, don't often arise. Many isolated communities have welcomed anthropologists; others have taken heads. Kwame Anthony Appiah (2005) has done some very important recent work which gives hope to those of us who fear both "tribalism" of the morally dangerous sort *and* the homogenization of the Earth by Western (especially American) culture. Appiah sees the ubiquity of television and cell phones as having sown the seeds of a potentially healthy "cosmopolitanism," with increasing numbers of people, rich and poor, seeing themselves as global citizens. At the same time, he sees little undermining of tradition, ritualistic practices, myths, and specialized moral conceptions.

22. I call it 'Jesusism' because most Christian churches do not endorse Jesus' message truthfully. Jainism could be added, but I focus on these three because they each have a large number of devotees. But consider Jainism in the mix. There may be others.

23. In ancient Chinese philosophy, especially on one reading of Confucius's *Analects* (Sarkissian, in progress), there is considerable emphasis on "moral charisma." The person who is *ren* (humane) is *de* (virtuous), and some very virtuous people have "moral charisma." Sometimes such charisma is characterized as engendering something like moral contagion: everyone catches the virtue of the charismatic person. But Confucius also raises concerns about what he calls "village worthies," charismatic types who produce the same effects, for example, gaining followers. Village worthies *seem* virtuous but really aren't. Shades of Thrasymachus and Machiavelli.

24. The reason attention is not normally paid to Buddhism and Jesusism by philosophers may have a simple explanation. They are conceived of as religious in the bad

sense, as asserting spooky non-sense. What Alasdair McIntyre calls the "project of the enlightenment" is to ground ethics on a secular foundation. The reason Kant's ethics does not make my list of ethical conceptions that endorse universal love is that it doesn't. Kant does endorse impartiality and he does speak of a kingdom of ends where everyone is treated as an end and never as a mere means. But Kant's ethics embodies a very powerful strain of the standard Christian construal of the "Golden Rule." That is, actively embodying the "golden rule," or the Buddhist's "four divine abodes," or being a dedicated consequentialist is *supererogatory*, above and beyond the call of duty. Jesus was not a Kantian, nor was the Buddha.

25. There is an underdiscussed debate in ancient Chinese philosophy between Confucius and Mencius, on one side and Mozi, on the other side. The first pair argue that *ren*, humaneness, needs to start in each family and then that it will spread. Mozi endorses a more expansive virtue, *jian ai*, universal love or solicitude, and is skeptical that without it, morality can escape some form of local chauvinism, village, nation-state, and so on.

26. One might say that what I call "simple empiricism" does not warrant giving an imprimatur of rationality to respect for logical consistency, because logical truths are not empirical. Kant, a rationalist, claimed that the demand of consistency was an a priori demand of Pure Reason. The naturalist denies that there is such a thing as Pure Reason. Here is a tactic that appeals to me. Over phylogenic time, consistency in belief has proved so adaptive that it is, as it were, an innate idea and/or disposition. It is *a priori* for that reason, in that sense. Being inconsistent got ancestors of humans in trouble. So we have a natural desire that is adaptive (indeed it evolved as an "adaptation" in the biological sense) to be consistent. Inadvertent inconsistency still gets us in trouble. Thus, it is rational to seek consistency.

27. Two problems for evolutionary psychology: (1) Why do 3–4-year-old boys "know" the names of dinosaurs? (2) Why are 18–19-year-old males extraordinarily prone to become Ayn Randians? Notice both pass after a certain temporally circumscribed critical stage. Arguably this is unfortunate in the first case but good in the second case.

28. I have a deep-seated conviction, which I don't know how to prove, that the monastic traditions of Buddhism and Christianity were in part designed to help with the natural problem of family chauvinism. When I was in Dharamsala meeting with the Dalai Lama in 2000, I mentioned to my son Ben (then 18 years old) how striking was the "warm glow" on the faces of the abundant monks and nuns. He commented: "Dad, they have no spouses and no teenage children."

Bibliography

Primary Buddhist Sources Consulted or Cited

Abhidhammattha Sangaha: A Comprehensive Manual of Abhidhamma. The Philosophical Psychology of Buddhism, ed. Bhikkhu Bodhi. BPS Pariyatti Editions, 2000.

Abhidhamma Studies: Buddhist Explorations of Consciousness and Time, ed. Ven. Nyanaponika Thera, Bhikkhu Bohdi, 1949 (Wisdom Publications, 1998).

Mind Training: The Great Collection, compiled by Shonu Gyalchok and Konchok Gyaltsen. Wisdom Publications, 2006.

Shantideva, *The Way of the Bodhisattva* (*Bodhicharyavatara*). Shambala, 1997.

The Connected Discourses of the Buddha. Wisdom Publications, 2000. Cited as SN.

The Long Discourses of the Buddha. Wisdom Publications, 1987 and 1995. Cited as DN.

The Middle Length Discourses of the Buddha. Wisdom Publications, 1995. Cited as MN.

General References

Abraham, A., et al. 2005. "Conceptual expansion and creative imagery as a function of psychoticism." *Consciousness and Cognition* 14: 520–534.

Abu-Lughod, L. 1986. *Veiled Sentiments: Honor and Poetry in a Bedouin Society*. University of California Press.

Abu-Lughod, L. 2001. *Dramas of Nationhood: The Politics of Television in Egypt*. University of Chicago Press.

Allport, G. 1943. *Becoming: Basic Considerations for a Psychology of Personality*. Yale University Press.

Annas, J. 1993. *The Morality of Happiness*. Oxford University Press.

Appiah, K. 1992. *In My Father's House: Africa in the Philosophy of Culture*. Oxford University Press.

Appiah, K. 2005. *The Ethics of Identity*. Princeton University Press.

Appiah, K. 2006. *Cosmopolitanism: Ethics in a World of Strangers*. Norton.

Arbib, M., and M. Hesse. 1986. *The Construction of Reality*. Cambridge University Press.

Aristotle. 1985. *Nicomachean Ethics*. Hackett.

Armstrong, K. 2000. *The Battle for God: A History of Fundamentalism*. Random House.

Aspinwall, L. B., and U. M. Staudinger, eds. 2003. *A Psychology of Human Strengths: Fundamental Questions and Future Directions for a Positive Psychology*. American Psychological Association.

Austin, J. H. 1998. *Zen and the Brain*. MIT Press.

Austin, J. H. 2006. *Zen-Brain Reflections*. MIT Press.

Barbour, I. 1990. *Religion in an Age of Science: The Gifford Lectures*, volume 1. Harper.

Barbour, I. 2000. *When Science Meets Religion: Enemies, Strangers, or Partners?* Harper.

Barkow, J. H., L. Cosmides, and J. Tooby, eds. 1992. *The Adapted Mind*. Oxford University Press.

Barnes, J., ed. 1995. *The Cambridge Companion to Aristotle*. Cambridge University Press.

Barrett, J. 2000. "Exploring the natural foundations of religion." *Trends in Cognitive Science* 4: 29–34.

Batchelor, S. 1997. *Buddhism Without Beliefs: A Contemporary Guide to Awakening*. Riverhead Books.

Becker, L. 1998. *A New Stoicism*. Princeton University Press.

Begley, S. 2007. *Train Your Mind, Change Your Brain*. Ballantine Books.

Bell, R. 2002. *Understanding African Philosophy: A Cross-Cultural Approach to Classical and Contemporary Issues*. Routledge.

Belliotti, R. A. 2004. *Happiness Is Overrated*. Rowman & Littlefield.

Bennett, M. R., and P. M. S. Hacker. 2003. *Philosophical Foundations of Neuroscience*. Blackwell.

Bickle, J. 1998. *Psychoneural Reduction: The New Wave*. MIT Press.

Block, N., O. Flanagan, and G. Güzeldere, eds. 1997. *The Nature of Consciousness: Philosophical Debates*. MIT Press.

Bloom, P. 2004. *Descartes' Baby: How the Science of Child Development Explains What Makes Us Human*. Basic Books.

Bond, G. D. 1988. "The Arahant: Sainthood in Theravada Buddhism." In *Sainthood: Its Manifestations in World Religions*, ed. R. Kieckhefer and G. Bond. University of California Press.

Boyer, P. 1994. *The Naturalness of Religious Ideas: A Cognitive Theory of Religion*. University of California Press.

Boyer, P. 2001. *Religion Explained: The Evolutionary Origins of Religious Thought*. Basic Books.

Bratman, M. 1987. *Intention, Plans, and Practical Reason*. Harvard University Press.

Brown, W. S., N. Murphy, and H. Malony, eds. 1998. *Whatever Happened to the Soul? Scientific and Theological Portraits of Human Nature*. Fortress.

Buller, D. J. 2005. *Adapting Minds: Evolutionary Psychology and the Persistent Quest for Human Nature*. MIT Press.

Burtt, E. A., ed. 1955. *The Teachings of the Compassionate Buddha: Early Discourses, the Dhammapada, and Later Basic Writings*. New American Library.

Cabezon, J., ed. 1988. *The Bodhgaya Interviews*. Snow Lion.

Cacioppo, J., and G. Berntson, eds. 2004. *Essays in Social Neuroscience*. MIT Press.

Cacioppo, J., et al., eds. 2002. *Foundations in Social Neuroscience*. MIT Press.

Carrithers, M., S. Collins, and S. Lukes, eds. 1985. *The Category of the Person: Anthropology, Philosophy, History*. Cambridge University Press.

Chalmers, D. 1995. "Facing up to the problem of consciousness." *Journal of Consciousness Studies* 2, no. 3: 200–219.

Chalmers, D. 1996. *The Conscious Mind: In Search of a Fundamental Theory*. Oxford University Press.

Chisholm, R. 1964. "Human freedom and the self." Reprinted in *Free Will*, ed. G. Watson (Oxford University Press, 1982).

Chisholm, R. 1976. *Person and Object*. Open Court.

Churchland, P. M. 1984. *Matter and Consciousness: A Contemporary Introduction to the Philosophy of Mind*. MIT Press.

Churchland, P. M. 1989. *A Neurocomputational Perspective: The Nature of Mind and the Structure of Science*. MIT Press.

Churchland, P. M. 1995. *The Engine of Reason, the Seat of the Soul: A Philosophical Journey into the Brain*. MIT Press.

Churchland, P. M. 1996. "Flanagan on moral knowledge." In *The Churchlands and Their Critics*, ed. R. McCauley. Blackwell.

Churchland, P. S. 1986. *Neurophilosophy: Toward a Unified Science of the Mind/Brain*. MIT Press.

Clayton, P. 2004. *Mind and Emergence: From Quantum to Consciousness*. Oxford University Press.

Clayton, P., and Z. Simpson, eds. 2006. *The Oxford Handbook of Religion and Science*. Oxford University Press.

Collins, S. 1982. *Selfless Persons: Imagery and Thought in Theravada Buddhism*. Cambridge University Press.

Conze, E. 1959. *Buddhism: Its Essence and Development*. Harper.

Crick, F. 1994. *The Astonishing Hypothesis: The Scientific Search for the Soul*. Scribner.

Csikszentmihalyi, M. 1990. *Flow: The Psychology of Optimal Experience*. Harper Perennial.

Csikszentmihalyi, M. 1997. *Creativity: Flow and the Psychology of Discovery and Invention*. Harper Perennial.

Dalai Lama. 1962. *My Land and My People* (Warner Books, 1997).

Dalai Lama. 1990. *A Policy of Kindness*. Snow Lion.

Dalai Lama. 1991. *The Path to Bliss*. Snow Lion.

Dalai Lama. 1994. *A Flash of Lightning in the Dark of Night: A Guide to the Bodhisattva's Way of Life*. Shambhala.

Dalai Lama. 1995. *Commentary on the Thirty-Seven Practices of a Bodhisattva*. Library of Tibetan Works and Archives.

Dalai Lama. 1999. *Ethics for the New Millennium*. Riverhead Books.

Dalai Lama. 2005. *The Universe in a Single Atom: The Convergence of Science and Spirituality*. Morgan Road Books.

Dalai Lama and H. Cutler. 1998. *The Art of Happiness: A Handbook for Living*. Riverhead Books.

Damasio, A. 1994. *Descartes' Error: Emotion, Reason, and the Human Brain*. Avon Books.

Damasio, A. 1999. *The Feeling of What Happens: Body and Emotion in the Making of Consciousness*. Harcourt Brace.

Damasio, A. 2003. *Looking for Spinoza: Joy, Sorrow, and the Feeling Brain*. Harcourt.

Darwin, C. 1859. *On the Origin of Species*. Modern Library.

Darwin, C. 1871. *The Descent of Man*. Modern Library.

Darwin, C. 1872. *The Expression of the Emotions in Man and Animals* (third edition, Oxford University Press, 1998).

Davids, R., and W. Stede. 1993. *Pali-English Dictionary*. Montilal Bandarsidass.

Davidson, R. J. 2000. "Affective style, psychopathology and resilience: Brain mechanisms and plasticity." *American Psychologist* 55: 1196–1214.

Davidson, R. J., ed. 2002. *Handbook of Affective Sciences*. Oxford University Press.

Davidson, R. J. 2003. "Affective neuroscience and psychophysiology: Toward a synthesis." *Psychophysiology* 40: 655–665.

Davidson, R. J. 2004. "Well-being and affective style: Neural substrates and biobehavioral Correlates." *Philosophical Transactions of the Royal Society (London)* 359: 1395–1411.

Davidson, R. J., and A. Harrington, eds. 2002. *Visions of Compassion: Western Scientists and Tibetan Buddhists Examine Human Nature*. Oxford University Press.

Davidson, R. J., and K. Hugdahl, eds. 2002. *The Asymmetrical Brain*. MIT Press.

Davidson, R. J., and W. Irwin. 1999. "The Functional neuroanatomy of emotion and affective style." *Trends in Cognitive Sciences* 3, no. 1: 11–21.

Davidson, R. J., D. J. Goleman, and G. E. Schwartz, eds. 1976. "Attentional and affective concomitants of meditation: A cross-sectional study." *Journal of Abnormal Psychology* 85: 235–238.

Davidson, R. J., et al. 2003. "Alterations in brain and immune function produced by mindfulness meditation." *Psychosomatic Medicine* 65: 564–570.

Dawkins, R. 1976. *The Selfish Gene*. Oxford University Press.

Dawkins, R. 1982. *The Extended Phenotype*. San Francisco; Freeman.

Dawkins, R. 1986. *The Blind Watchmaker*. Norton.

Dawkins, R. 2006. *The God Delusion*. Houghton Mifflin.

Deacon, T. 1997. *The Symbolic Species: The Co-evolution of Language and the Brain*. Norton.

De Caro, M., and D. Macarthur, eds. 2004. *Naturalism in Question*. Harvard University Press.

Dennett, D. C. 1978. *Brainstorms: Philosophical Essays on Mind and Psychology*. MIT Press.

Dennett, D. C. 1984a. *Elbow Room: The Varieties of Free Will Worth Wanting*. MIT Press.

Dennett, D. C. 1984b. "I could not have done otherwise: So what?" *Journal of Philosophy* 81, no. 10: 553–565.

Dennett, D. C. 1988. "Why everyone is a novelist." *Times Literary Supplement* 4, no. 459: 1016–1022.

Dennett, D. C. 1991. *Consciousness Explained.* Little, Brown.

Dennett, D. C. 1995. *Darwin's Dangerous Idea: Evolution and the Meanings of Life.* Simon & Schuster.

Descartes, R. *Discourse on Method and Meditations on First Philosophy.* Hackett, 1993.

de Sousa, R. 1987. *The Rationality of Emotion.* MIT Press.

Devlin, P. 1965. *The Enforcement of Morals.* Oxford University Press.

De Waal, F. 2006. *Primates and Philosophers: How Morality Evolved.* Princeton University Press.

Dewey, J. 1894. "The ego as cause." *Philosophical Review* 3: 337–341.

Dewey, J. 1922. *Human Nature and Conduct.* Holt.

Dewey, J. 1934. *Art as Experience.* Perigee Books.

Diener, E., and E. M. Suh, eds. 2000. *Culture and Subjective Well-Being.* MIT Press.

Diener, E., and C. Scollon. 2003. "Subjective well-being is desirable but not the summum bonum." Conference paper, University of Illinois.

Drees, W. 1996. *Religion, Science and Naturalism.* Cambridge University Press.

Dretske, F. 1995. *Naturalizing the Mind.* MIT Press.

Dreyfus, G. 1997. *Recognizing Reality: Dharmakirti's Philosophy and Its Tibetan Interpretation.* SUNY Press.

Dreyfus, G. 2003. *The Sound of Two Hands Clapping: The Education of a Tibetan Buddhist Monk.* California.

Droit, R. P. 2003. *The Cult of Nothingness: The Philosophers and the Buddha.* University of North Carolina Press.

Dunne, J. 2004. *Foundations of Dharmakirti's Philosophy.* Wisdom Publications.

Easterlin, R. A. 2003. "Explaining happiness." In *Proceedings of the National Academy of Science: Inaugural Articles by members of the National Academy elected on April 30, 2002.*

Easterlin, R. A. 2004. Money, Sex, and Happiness: An Empirical Study. Working Paper 10499, National Bureau of Economic Research.

Edelman, G., and G. Tononi. 2000. *A Universe of Consciousness: How Matter Becomes Imagination*. Basic Books.

Einstein, G., and O. Flanagan. 2003. "Sexual identities and narratives of self." In *Narrative and Consciousness*, ed. G. Fireman et al., Oxford University Press.

Ekman, P. 1972. *Emotions in the Human Face*. Pergamon.

Ekman, P. 1992. "Are there basic emotions?" *Psychological Review* 99, no. 3: 550–553.

Ekman, P. 1998. Introduction, afterword, and commentaries. In C. Darwin, *The Expression of the Emotions in Man and Animals*, third edition, ed. P. Ekman. Oxford University Press.

Ekman, P. 2003. *Emotions Revealed: Recognizing Faces and Feelings to Improve Communication and Emotional Life*. Holt/Times Books.

Ekman, P., J. Campos., R. J. Davidson, and F. De Waals. 2003. *Emotions Inside Out. Annals of the New York Academy of Sciences* 1000.

Ekman, P., R. J. Davidson, M. Ricard, and A. Wallace. 2005. "Buddhist and psychological perspectives on emotions and well-being." *Current Directions in Psychological Science* 14: 59–63.

Ekman, P., R. W. Levinson, and W. V. Freisen. 1985. "Autonomic nervous system activity distinguished among emotions." *Science* 22: 1208–1210.

Erikson, E. 1950. *Childhood and Society*, second edition. Norton.

Eze, E., ed. 1998. *African Philosophy: An Anthology*. Blackwell.

Fesmire, S. 2003. *John Dewey and Moral Imagination: Pragmatism in Ethics*. Indiana University Press.

Festinger, L. 1957. *A Theory of Cognitive Dissonance*. Row, Peterson.

Fetzer, J., ed. 2002. *Evolving Consciousness*. John Benjamins.

Fireman, G. D., T. E. McVay, and O. Flanagan, eds. 2003. *Narrative and Consciousness: Literature, Psychology, and the Brain*. Oxford University Press.

Flanagan, O. 1984. *The Science of the Mind*. MIT Press.

Flanagan, O. 1991a. *The Science of the Mind*, second edition. MIT Press.

Flanagan, O. 1991b. *Varieties of Moral Personality: Ethics and Psychological Realism*. Harvard University Press.

Flanagan, O. 1992. *Consciousness Reconsidered*. MIT Press.

Flanagan, O. 1996a. *Self Expressions: Mind, Morals, and the Meaning of Life*. Oxford University Press.

Flanagan, O. 1996b. "The moral network." In *The Churchlands and Their Critics*, ed. R. McCauley. Blackwell.

Flanagan, O. 2000a. "Destructive emotions." *Consciousness and Emotion* 1, no. 2: 67–88, 259–281.

Flanagan, O. 2000b. *Dreaming Souls: Sleep, Dreams, and the Evolution of the Conscious Mind*. Oxford University Press.

Flanagan, O. 2002. *The Problem of the Soul: Two Visions of Mind and How to Reconcile Them*. Basic Books.

Flanagan, O. 2003a. "The colour of happiness." *New Scientist* 178, no. 2396: 44.

Flanagan, O. 2003b. "Ethical expressions: Why moralists scowl, frown, and smile." In *The Cambridge Companion to Darwin*. Cambridge University Press.

Flanagan, O. 2006a. "The Bodhisattva's brain: Neuroscience and happiness." In *Buddhist Thought and Applied Psychological Research: Transcending the Boundaries*, ed. D. Nauriyal et al. Routledge.

Flanagan, O. 2006b. "Varieties of naturalism: The many meanings of naturalism." In *The Oxford Handbook of Religion and Science*, ed. P. Clayton and Z. Simpson. Oxford University Press.

Flanagan, O., and T. Polger. 1995. "Zombies and the function of consciousness." *Journal of Consciousness Studies* 2, no. 4: 313–321.

Flanagan, O., and A. O. Rorty, eds. 1990. *Identity, Character, and Morality: Essays in Moral Psychology*. MIT Press.

Flanagan, O., H. Sarkissian, and D. Wong. 2007a. "Naturalizing ethics." In *Moral Psychology*, volume 1: *The Evolution of Morality: Adaptations and Innateness*, ed. W. Sinnott-Armstrong. MIT Press.

Flanagan, O., H. Sarkissian, and D. Wong. 2007b. "What is the nature of morality? A response to Casebeer, Railton, and Ruse." In *Moral Psychology*, volume 1: *The Evolution of Morality: Adaptations and Innateness*, ed. W. Sinnott-Armstrong. MIT Press.

Foot, P. 1972. "Morality as a system of hypothetical imperatives." Reprinted in Foot, *Virtues and Vices* (Clarendon, 1978).

Foot, P. 2001. *Natural Goodness*. Clarendon.

Frank, R. 2004. "How not to buy happiness," *Daedalus* 133, no. 2: 69–79.

Frank, S. 1998. *Foundations of Social Evolution*. Princeton University Press.

Frankfurt, H. 1988. *The Importance of What We Care About*. Cambridge University Press.

Frankfurt, H. 2004. *The Reasons of Love*. Princeton University Press.

Frankfurt, H. 2005. *On Bullshit*. Princeton University Press.

Freud, S. 1927. *The Future of an Illusion* (Doubleday/Anchor, 1961).

Frey, B., and A. Stutzer. 2002. *Happiness and Economics: How the Economics and Institutions Affect Well-Being*. Princeton University Press.

Frith, C., and D. Wolpert, eds. 2003. *The Neuroscience of Social Interaction: Decoding, Imitating, and Influencing the Actions of Others*. Oxford University Press.

Fromm, E. 1955. *The Sane Society*. Rinehart.

Gallagher, S. 2005. *How the Body Shapes the Mind*. Oxford University Press.

Gardner, H. 1985. *The Mind's New Science: A History of Cognitive Revolution*. Basic Books.

Garfield, J., tr. 1995. *The Fundamental Wisdom of the Middle Way: Nagarjuna's Mulamadhyamakakarika*. Oxford University Press.

Garfield, J. 2002. *Empty Words: Buddhist Philosophy and Cross-Cultural Interpretation*. Oxford University Press.

Gazzaniga, M., ed. 1995. *The Cognitive Neurosciences*. MIT Press.

Geertz, C. 1973. *The Interpretation of Cultures: Selected Essays*. Basic Books.

Gerth, H., and C. W. Mills, eds. 1946. *From Max Weber: Essays in Sociology*. Oxford University Press.

Gibbard, A. 1990. *Wise Choices, Apt Feelings: A Theory of Normative Judgment*. Harvard University Press.

Gibbard, A. 2003. *Thinking How to Live*. Harvard University Press.

Goldman, A. 1986. *Epistemology and Cognition*. Harvard University Press.

Goleman, D. 2003a. *Destructive Emotions: How Can We Overcome Them? A Scientific Dialogue with the Dalai Lama*. Bantam Books.

Goleman, D. 2003b. "Finding happiness: Cajole your brain to lean left." *New York Times*, February 4.

Goleman, D. 2006. *Social Intelligence: The New Science of Human Relationships*. Bantam.

Gombrich, R. F. 2002. *How Buddhism Began: The Conditioned Genesis of the Early Teachings*. Munshiram Manoharlal.

Goodenough, U. 1998. *The Sacred Depths of Nature*. Oxford University Press.

Goodman, C. 2007. Consequences of Compassion: An Interpretation and Defense of Buddhist Ethics. Manuscript.

Goodman, N. 1976. *Languages of Art: An Approach to a Theory of Symbols*. Hackett.

Goodman, N. 1978. *Ways of Worldmaking*. Hackett.

Gouda, F. 1995. *Poverty and Political Culture: The Rhetoric of Social Welfare in the Netherlands and France, 1815–1854*. Rowman & Littlefield.

Gould, S. J. 1999. *Rocks of Ages: Science and Religion in the Fullness of Life*. Ballantine.

Greene, J. 2002. The Terrible, Horrible, No good, Very Bad Truth About Morality and What To Do About It. Ph.D. dissertation, Princeton University.

Greene, J., and J. Haidt. 2002. "How (and where) does moral judgment work?" *Trends in Cognitive Sciences* 6, no. 12: 517–523.

Greene, J., R. B. Sommerville, L. E. Nystrom, J. M. Darley, and J. D. Cohen. 2001. "An fMRI investigation of emotional engagement in moral judgment." *Science*: 293, September 14: 2105–2108.

Griffiths, P. E. 1997. *What Emotions Really Are: The Problem of Psychological Categories*. University of Chicago Press.

Grimes, J. 1989. *A Concise Dictionary of Indian Philosophy: Sanskrit Terms Defined in English*. State University of New York Press.

Güzeldere, G. 1997. "The many faces of consciousness: A field guide." In *The Nature of Consciousness: Philosophical Debates*, ed. N. Block et al. MIT Press.

Gyatso, Tenzin. 2003a. "The Monk in the Lab." *New York Times*, April 2.

Gyatso, Tenzin. 2003b. "On the luminosity of being." *New Scientist* 178, no. 2396: 42–43.

Hahn, Thich Naht. 1987. *Being Peace*. Parallax.

Hahn, Thich Naht. 2004. *Joyfully Together: The Art of Building a Harmonious Community*. Parallax.

Haidt, J. 2001. "The emotional dog and its rational tail: A social intuitionist approach to moral judgment." *Psychological Review* 108: 814–834.

Haidt, J. 2006. *The Happiness Hypothesis: Finding Modern Truth in Ancient Wisdom*. Basic Books.

Haidt, J., and C. Joseph. 2004. "Intuitive ethics: How innately prepared intuitions generate culturally variable virtues." *Daedalus*, Special Issue on Human Nature: 55–66.

Hamer, D. 2004. *The God Gene: How Faith Is Hardwired into Our Genes*. Anchor Books.

Harris, S. 2004. *The End of Faith: Religion, Terror, and the Future of Reason*. Norton.

Harris, S. 2006. *Letter to a Christian Nation*. Knopf.

Harvey, P. 1990. *An Introduction to Buddhism*. Cambridge University Press.

Harvey, P. 2000. *An Introduction to Buddhist Ethics*. Cambridge University Press.

Hauser, M. 2006. *Moral Minds: How Nature Designed Our Universal Sense of Right and Wrong*. HarperCollins.

Heidegger, M. 1962. *Being and Time*. Harper & Row.

Heine, S. J., et al. 1999. "Is there a universal need for positive self-regard?" *Psychological Review* 106, no. 4: 766–794.

Henrich, J., et al., eds. 2004. *Foundations of Human Sociality: Economic Experiments and Ethnographic Evidence from Fifteen Small-Scale Societies*. Oxford University Press.

Herman, D., ed. 2003. *Narrative Theory and the Cognitive Sciences*. CSLI.

Hobbes, Thomas. 1651. *Leviathan* (Clarendon, 1999).

Hodge, J., and G. Radick., eds. 2003. *The Cambridge Companion to Darwin*. Cambridge University Press.

Hoogcarspel, E. 2005. *Nagarjuna: The Central Philosophy, Basic Verses*. Olive.

Hopkins, J. 1996. *Meditation on Emptiness*. Wisdom Publications.

Hutchinson, D. 1995. "Ethics." In *The Cambridge Companion to Aristotle*. Cambridge University Press.

Hume, D. 1739. *Treatise on Human Nature* (Penguin, 1985).

Hume, D. 1777a. *An Enquiry Concerning Human Understanding* (Hackett, 1977).

Hume, D. 1777b. *An Enquiry Concerning the Principles of Morals* (Hackett, 1983).

Husserl, E. 1913. *Ideas* (Collier, 1972).

Imbo, S. 1998. *An Introduction to African Philosophy*. Rowman & Littlefield.

Ivanhoe, P. J. 2000. *Confucian Moral Self Cultivation*. Hackett.

Jahoda, M. 1953. "The meaning of psychological health." *Social Casework* 34: 349–354.

Jahoda, M. 1958. *Current Concepts of Positive Mental Health*. Basic Books.

Jain, S. K., ed. 1997. *Glimpses of Jainism*. Motilal Banarsidass.

James, W. 1890. *Principles of Psychology*. Holt.

James, W. 1892. *Psychology: The Briefer Course*. Harper.

James, W. 1897. *The Will To Believe and Other Essays in Popular Philosophy*. Dover.

James, W. 1898. *The Varieties of Religious Experience* (Penguin, 1982).

Jamison, K. R. 1993. *Touched With Fire: Manic-Depressive Illness and the Artistic Temperament.* Free Press.

Jamison, K. R. 1995. *An Unquiet Mind: A Memoir of Moods and Madness*. Vintage Books.

Jamison, K. R. 2004. *Exuberance: The Passion For Life*. Knopf.

Jhingran, S. 1989. *Aspects of Hindu Morality*. Motilal Banarsidass.

Jinpa, Thupten. 2003. "Sciene as an ally or a rival philosophy? Tibetan Buddhist thinkers' engagement with modern science." In *Buddhism and Science: Breaking New Ground*, ed. A. Wallace. Columbia University Press.

Kabat-Zinn, J. 1994. *Wherever You Go There You Are: Mindfulness Meditation in Everyday Life*. Hyperion.

Kahneman, D. 1999. "Objective happiness." In *Well-Being: The Foundations of Hedonic Psychology*, ed. D. Kahneman et al. Russell Sage Foundation.

Kahneman, D., E. Diener, and N. Schwarz, eds. 1999. *Well-Being: The Foundations of Hedonic Psychology*. Russell Sage Foundation.

Kalupahana, D. 1992. *A History of Buddhist Philosophy: Continuities and Discontinuities*. University of Hawaii Press.

Kapstein, M. 2001. *Reason's Traces: Identity and Interpretation in Indian and Tibetan Buddhist Thought.* Wisdom Publications.

Kauffman, S. 1995. *At Home in the Universe: The Search for the Laws of Self-Organization and Complexity*. Oxford University Press.

Keown, D. 1992. *The Nature of Buddhist Ethics* (Palgrave, 2001).

Keown, D., ed. 2000. *Contemporary Buddhist Ethics*. Curzon.

Keown, D. 2003. *Dictionary of Buddhism*. Oxford University Press.

Keyes, C., and J. Haidt, eds. 2003. *Flourishing: Positive Psychology and the Life Well-Lived*. American Psychological Association.

Kieckhefer, R., and G. D. Bond, eds. 1988. *Sainthood: Its Manifestations in World Religions*. University of California Press.

Kim, J. 1993. *Supervenience and Mind: Selected Philosophical Essays*. Cambridge University Press.

Kim, J. 2005. *Physicalism, or Something Near Enough*. Princeton University Press.

Kim, J. 2006. *The Philosophy of Mind*, second edition. Westview.

Koch, C. 2004. *The Quest for Consciousness: A Neurobiological Approach*. Roberts.

Koonz, C. 1987. *Mothers in the Fatherland: Women, the Family, and Nazi Politics*. St. Martin's Press.

Kuhn, T. 1957. *The Copernican Revolution: Planetary Astronomy in the Development of Western Thought*. Harvard University Press.

Kuhn, T. 1962. *The Structure of Scientific Revolutions* (second edition: University of Chicago Press, 1970).

Kupperman, J. 1991. *Character*. Oxford University Press.

Kurzweil, R. 2004. *The Singularity Is Near: When Humans Transcend Biology*. Viking.

Lakoff, G. 1966. *Moral Politics: What Conservatives Know That Liberals Don't*. University of Chicago Press.

Lane, R., and L. Nadel, eds. 2000. *Cognitive Neuroscience of Emotion*. Oxford University Press.

Lawson, E. T., and R. N. McCauley. 1990. *Rethinking Religion: Connecting Cognition and Culture*. Cambridge University Press.

LeDoux, J. 1996. *The Emotional Brain: The Mysterious Underpinnings of Emotional Life*. Touchstone Books.

Lekuton, J. 2003. *Facing the Lion: Growing Up Maasai on the African Savanna*. National Geographic Society.

Levine, J. 1983. "Materialism and qualia: The explanatory gap." *Pacific Philosophical Quarterly* 64: 345–361.

Levine, J. 2001. *Purple Haze: The Puzzle of Consciousness*. Oxford University Press.

Logothetis, N., and J. D. Schall. 1989. "Neuronal correlates of subjective visual perception." *Science* 245: 761–763.

Lopez, D. 1988. "Sanctification on the Bodhisattva path." In *Sainthood: Its Manifestations in World Religions*, ed. R. Kieckhefer and G. Bond. University of California Press.

Lopez, D. 2001. *The Story of Buddhism: A Concise Guide to its History and Teachings*. HarperCollins.

Lopez, D. 2006. *The Madman's Middle Way. Reflections on Reality of the Tibetan Monk Gendun Chöpel*. University of Chicago Press.

Lopez, S. J., and C. R. Snyder, eds. 2003. *Positive Psychological Assessment: A Handbook of Models and Measures*. American Psychological Association.

Lutz, A., L. Greischar, N. Rawlings, M. Ricard, and R. Davidson. 2004. "Long-term mediators self-induce high-amplitude gamma synchrony during mental practice." *Proceedings of the National Academy of Science* 101, no. 46: 16369–16373.

MacIntyre, A. 1981. *After Virtue*. University of Notre Dame Press.

MacIntyre, A. 1988. *Whose Justice? Which Rationality?* University of Notre Dame Press.

MacIntyre, A. 1990. *Three Rival Versions of Moral Enquiry*. University of Notre Dame Press.

MacIntyre, A. 1999. *Dependent Rational Animals*. Carus.

Mahapragya, A. 2000. *Jainism and Its Philosophical Foundations*. Anmol.

Marcel, A., and E. Basiach. 1988. *Consciousness in Contemporary Science*. Clarendon.

Marx, K. *The Economic and Philosophical Manuscripts of 1844* (International Publishers, 1975).

Masolo, D. 1994. *African Philosophy in Search of Identity*. Bloomington and Indiana University Press.

Maturana, H., and F. Varela. 1998. *The Tree of Knowledge: The Biological Roots of Human Understanding*. Shambhala.

May, L., M. Friedman, and A. Clark, eds. 1996. *Mind and Morals: Essays on Ethics and Cognitive Science*. MIT Press.

Mayr, E. 2001. *What Evolution Is*. Basic Books.

McCauley, R., ed. 1996. *The Churchlands and Their Critics*. Blackwell.

McCauley, R. N., and E. T. Lawson. 2002. *Bringing Ritual to Mind: Psychological Foundations of Cultural Forms*. Cambridge University Press.

McDowell, J. 1994. *Mind and World*. New edition, Harvard University Press, 1996.

McDowell, J. 2004. "Naturalism in the philosophy of mind." In *Naturalism in Question*, ed. M. De Caro and D. Macarthur. Harvard University Press.

McEvilley, T. 2002. *The Shape of Ancient Thought: Comparative Studies in Greek and Indian Philosophies*. Allworth.

McGinn, C. 1989. "Can we solve the mind-body problem?" *Mind* 97, no. 891: 349–366.

McGinn, C. 1991. *The Problem of Consciousness*. Blackwell.

McGinn, C. 1999. *The Mysterious Flame: Conscious Minds in a Material World*. Basic Books.

McMahon, D. 2006. *Happiness: A History*. Atlantic Monthly Press.

Menninger, K. 1930. "What is a healthy mind?" In *The Healthy-Minded Child*, ed. N. Crawford and K. Menninger. Coward-McCann.

Merleau-Ponty, M. 1962. *Phenomenology of Perception*. Humanities Press.

Metzinger, T., ed. 2000. *Neural Correlates of Consciousness: Empirical and Conceptual Questions*. MIT Press.

Miller, G., E. Galanter, and K. Pribram. 1960. *Plans and the Structure of Behavior*. Holt.

Moran, R. 2001. *Authority and Estrangement: An Essay on Self-Knowledge*. Princeton University Press.

Mozi. 2003. *Basic Writings*. Columbia University Press.

Mudimbe, V. 1988. *The Invention of Africa: Gnosis, Philosophy, and the Order of Knowledge*. Indiana University Press.

Myers, D. G. 1992. *The Pursuit of Happiness: Discovering the Pathway to Fulfillment, Well-Being, and Enduring Personal Joy*. Harper.

Nagel, T. 1979a. "What is it like to be a bat?" In *Mortal Questions*. Cambridge University Press.

Nagel, T. 1979b. "Moral luck." In *Mortal Questions*. Cambridge University Press.

Nauriyal, D. K., M. S. Drummond, and Y. B. Lal, eds. 2006. *Buddhist Thought and Applied Psychological Research: Transcending the Boundaries*. Routledge.

Newberg, A., E. D'Aquili, and V. Rause. *Why God Won't Go Away: Brain Science and The Biology of Belief*. Ballantine.

Nichols, S. 2004. *Sentimental Rules: On the Natural Foundations of Moral Judgment*. Oxford University Press.

Nichols, S., and S. P. Stich. 2003. *Mindreading: An Integrated Account of Pretence, Self-Awareness, and Understanding Other Minds*. Clarendon.

Nöe, A. 2004. *Action in Perception*. MIT Press.

Nordenstam, T. 1968. *Sudanese Ethics*. Scandinavian Institute of African Studies.

Nussbaum, M. 1986. *The Fragility of Goodness: Luck and Ethics in Greek Tragedy and Philosophy*. Cambridge University Press.

Nussbaum, M. 1994. *The Therapy of Desire: Theory and Practice in Hellenistic Ethics*. Princeton University Press.

Nussbaum, M. 1999. *Sex and Social Justice*. Oxford University Press.

Nussbaum, M. 2000. *Women and Human Development: The Capabilities Approach*. Cambridge University Press.

Nussbaum, M., and A. Sen, eds. 1992. *The Quality of Life*. Oxford University Press.

Obeyesekere, G. 2002. *Imagining Karma: Ethical Transformation in Amerindian, Buddhist, and Greek Rebirth.* University of California Press.

Pandita, S. 1992. *In This Very Life: Liberation Teachings of the Buddha.* Wisdom Publications.

Panksepp, J. 1998. *Affective Neuroscience: The Foundations of Human and Animal Emotions.* Oxford University Press.

Perry, J. 2001. *Knowledge, Possibility, and Consciousness.* MIT Press.

Peterson, C., and M. E. P. Seligman. 2004. *Character Strengths and Virtues: A Handbook and Classification.* Oxford University Press.

Petitot, J. et al., eds. 1999. *Naturalizing Phenomenology: Issues in Contemporary Phenomenology and Cognitive Science.* Stanford University Press.

Pincoffs, E. L. 1986. *Quandaries and Virtues: Against Reductivism in Ethics.* University Press of Kansas.

Pinker, S. 1995. *The Language Instinct.* Harper Collins.

Pinker, S. 1997. *How the Mind Works.* Norton.

Pockett, S., W. Banks, and S. Gallagher, eds. 2006. *Does Consciousness Cause Behavior?* MIT Press.

Polger, T. W. 2004. *Natural Minds.* MIT Press.

Polger, T., and O. Flanagan. 2002. "Consciousness, adaptation, and epiphenomenalism." In *Evolving Consciousness,* ed. J. Fetzer. John Benjamins.

Polkinghorne, J. 1998. *Belief in God in an Age of Science.* Yale University Press.

Polkinghorne, J. 2005. *Exploring Reality: The Intertwining of Science and Religion.* Yale University Press.

Post, S. 2004. Religion & Spirituality and Human Flourishing. Field Analysis. www.metanexus.net/tarp.

Purves, D., et al. 1997. *Neuroscience.* Sinauer.

Putnam, H. 1981. *Reason, Truth and History.* Cambridge University Press.

Putnam, H. 1999. *The Threefold Cord: Mind, Body, and World.* Columbia University Press.

Putnam, H. 2002. *The Collapse of the Fact/Value Dichotomy, and Other Essays.* Harvard University Press.

Putnam, H. 2004. *Ethics Without Ontology.* Harvard University Press.

Rahula, W. 1959. *What the Buddha Taught.* Grove.

Rawls, J. 1971. *Theory of Justice*. Harvard University Press.

Revonsuo, A. 2006. *Inner Presence: Consciousness as a Biological Phenomenon*. MIT Press.

Ricard, M. 2006. *Happiness: A Guide to Developing Life's Most Important Skill*. Little, Brown.

Rinchen, G. 1998. *The Six Perfections*. Snow Lion.

Rinchen, G. 2006. *How Karma Works: The Twelve Links of Dependent Arising*. Snow Lion.

Roland, A. 1988. *In Search of Self in India and Japan: Toward a Cross-Cultural Psychology*. Princeton University Press.

Rorty, R. 1979. *Philosophy and the Mirror of Nature*. Princeton University Press.

Rosenkrantz, M., D. Jackson, K. Dalton, I. Dolski, C. Ryff, B. Singer, D. Muller, N. Kalin, and R. Davidson. 2003. "Affective style and in vivo immune response: Neurobehavioral mechanisms." *Proceedings of the National Academy of Sciences* 100: 11148–11152.

Rothberg, D. 2006. *The Engaged Spiritual Life: A Buddhist Approach to Transforming Ourselves and the World*. Beacon.

Ruse, M. 2001. *Can a Darwinian Be a Christian?* Cambridge University Press.

Russell, R. J., N. Murphy, T. C. Meyering, and M. A. Arbib, eds. 1999. *Neuroscience and the Person: Scientific Perspectives on Divine Action*. Vatican Observatory Publications and Center for Theology and the Natural Sciences.

Ryff, C. 1989. "Happiness is everything or is it?" *Journal of Personal and Social Psychology* 57: 1069–1081.

Ryff, C., and Keyes, C. L. M. 1995. "The structure of psychological well-being revisited." *Journal of Personal and Social Psychology* 69: 719–727.

Ryle, G. 1949. *The Concept of Mind*. University of Chicago Press.

Saddhatissa, H. 1970. *Buddhist Ethics* (Wisdom Publications, 1997, 2003).

Saitoti, Tepilit Ole. 1986. *The Worlds of a Maasai Warrior; An Autobiography*. University of California Press.

Sarkissian, H. In progress. After Confucius: Empirical Psychology and Moral Power. Ph.D thesis, Duke University.

Seligman, M. 2002. *Authentic Happiness: Using the New Positive Psychology to Realize Your Potential for Lasting Fulfillment*. Free Press.

Sellars, W. 1963. "Philosophy and the scientific image of man." In Sellars, *Science, Perception, and Reality*. Humanities Press.

Sen, A. 1980. "Equality of what?" In *Tanner Lectures on Human Values*, volume 1, ed. S. McCurrin. Cambridge University Press.

Sen, A. 1987. *On Ethics and Economics*. Blackwell.

Shallice, T. 1988. *From Neuropsychology to Mental Structure*. Cambridge University Press.

Shun, K., and D. Wong, eds. 2004. *Confucian Ethics: A Comparative Study of Self, Autonomy, and Community*. Cambridge University Press.

Shweder, R., and R. LeVine, eds. 1984. *Culture Theory: Essays on Mind, Self, and Emotion*. Cambridge University Press.

Siderits, M. 2003. *Personal Identity and Buddhist Philosophy: Empty Persons*. Ashgate.

Siderits, M. 2007. *Buddhism and Philosophy*. Hackett.

Slingerland, E. 2003. *Effortless Action: Wu-Wei as Conceptual Metaphor and Spiritual Ideal in Early China*. Oxford University Press.

Snow, C. P. 1959. *The Two Cultures*. Cambridge University Press.

Snyder, C. R., and S. J. Lopez, eds. 2005. *Handbook of Positive Psychology*. Oxford University Press.

Sober, E., and D. S. Wilson. 1998. *Unto Others: The Evolution and Psychology of Unselfish Behavior*. Harvard University Press.

Sorabji, R. 2006. *Self: Ancient and Modern Insights about Individuality, Life, and Death*. University of Chicago Press.

Spinoza, Benedict de. *Ethics*. Oxford University Press, 2000.

Sterelny, K. 2003. *Thought in a Hostile World: The Evolution of Human Cognition*. Blackwell.

Sterelny, K., and P. E. Griffiths. 1999. *Sex and Death: An Introduction to Philosophy of Biology*. University of Chicago Press.

Strawson, G. 1994. *Mental Reality*. MIT Press.

Strawson, P. F. 1962. "Freedom and resentment." Reprinted in *Free Will*, ed. G. Watson (Oxford University Press, 1982).

Stroud, B. 1996. "The charm of naturalism." *Proceedings and Addresses of the American Philosophical Association* 70: 43–55.

Sutin, L. 2006. *All is Change: The Two-Thousand-Year Journey of Buddhism to the West*. Little, Brown.

Tancredi, L. 2005. *Hardwired Behavior: What Neuroscience Reveals about Morality*. Cambridge University Press.

Taylor, C. 2002. *Varieties of Religion Today: William James Revisited*. Harvard University Press.

Taylor, S. E. 1989. *Positive Illusions: Creative Self-Deception and the Healthy Mind*. Basic Books.

Taylor, S. E. 2002. *The Tending Instinct: How Nurturing is Essential to Who We Are and How We Live*. Time Books.

Taylor, S. E., and J. Brown. 1988. "Illusion and well-being: A social psychological perspective on mental health." *Psychological Bulletin* 103: 193–210.

Thompson, E. 2007. *Mind and Life*. Harvard University Press.

Thurman, R., and S. Batchelor. 1997. "Reincarnation: A debate." *Tricycle* 24, summer: 24–27, 109–116.

Tye, M. 2003. *Consciousness and Persons: Unity and Identity*. MIT Press.

Varela, F. 1999a. "Neurophenomenology." In *Naturalizing Phenomenology*, ed. J. Petitot et al. Stanford University Press.

Varela, F. 1999b. *Ethical Know-How: Action, Wisdom, and Cognition*. Stanford University Press.

Varela, F., E. Thompson, and E. Rosch. 1991. *The Embodied Mind: Cognitive Science and Human Experience*. MIT Press.

Wallace, A. 1993. *Tibetan Buddhism From the Ground Up*. Wisdom Publications.

Wallace, A. 2000. *The Taboo of Subjectivity: Toward a New Science of Consciousness*. Oxford University Press.

Wallace, A. 2003. *Buddhism and Science: Breaking New Ground*. Columbia University Press.

Wallace, A. 2005. *Genuine Happiness: Meditation as the Path to Fulfillment*. Wiley.

Wallace, A. 2006. *The Attention Revolution: Unlocking the Power of the Focused Mind*. Wisdom Publications.

Wallace, A. 2007. *Contemplative Science: Where Buddhism and Neuroscience Converge*. Columbia University Press.

Wattles, J. 1996. *The Golden Rule*. Oxford University Press.

Weber, M. 2004. "Science as a vocation." In *The Vocation Lectures*, ed. D. Owen and T. Strong. Hackett.

Whitehouse, H. 2004. *Modes of Religiosity: A Cognitive Theory of Religious Transmission.* Altamira.

Williams, B. 1985. *Ethics and the Limits of Philosophy*. Harvard University Press.

Williams, P. 1989. *Mahayana Buddhism*. Routledge.

Willson, M. 1987. *Rebirth and Western Buddhism*. Wisdom Publications.

Wilson, D. S. 2002. *Darwin's Cathedral: Evolution, Religion, and the Nature of Society*. University of Chicago Press.

Wilson, E. O. 1975. *Sociobiology: The New Synthesis*. Harvard University Press.

Wilson, E. O. 1978. *On Human Nature*. Harvard University Press.

Wilson, T. D. 2002. *Strangers to Ourselves: Discovering the Adaptive Unconscious*. Belknap.

Wollheim, R. 1999. *On the Emotions*. Yale University Press.

Wong, D. 1984. *Moral Relativity*. University of California Press.

Wong, D. 2006. *Natural Moralities: A Defense of Pluralistic Relativism*. Oxford University Press.

Wright, R. 1994. *The Moral Animal: The New Science Of Evolutionary Psychology*. Random House.

Wright, R. 2000. *Nonzero: The Logic of Human Destiny*. Vintage Books.

Zahavi, D. 2005. *Subjectivity and Selfhood: Investigating the First-Person Perspective*. MIT Press.

Index